ROUTLEDGE LIBRARY EDITIONS:
NUCLEAR SECURITY

Volume 5

INTERNATIONALIZATION TO PREVENT THE SPREAD OF NUCLEAR WEAPONS

INTERNATIONALIZATION TO PREVENT THE SPREAD OF NUCLEAR WEAPONS

STOCKHOLM INTERNATIONAL PEACE RESEARCH INSTITUTE

Routledge
Taylor & Francis Group
LONDON AND NEW YORK

First published in 1980 by Taylor & Francis Ltd

This edition first published in 2021
by Routledge
2 Park Square, Milton Park, Abingdon, Oxon OX14 4RN

and by Routledge
605 Third Avenue, New York, NY 10017

Routledge is an imprint of the Taylor & Francis Group, an informa business

British Library Cataloguing in Publication Data
A catalogue record for this book is available from the British Library

ISBN: 978-0-367-50682-7 (Set)
ISBN: 978-1-00-309763-1 (Set) (ebk)
ISBN: 978-0-367-50966-8 (Volume 5) (hbk)
ISBN: 978-1-00-305200-5 (Volume 5) (ebk)

Publisher's Note
The publisher has gone to great lengths to ensure the quality of this reprint but points out that some imperfections in the original copies may be apparent.

Disclaimer
The publisher has made every effort to trace copyright holders and would welcome correspondence from those they have been unable to trace.

Internationalization
to Prevent the Spread of Nuclear Weapons

sipri

Stockholm International Peace Research Institute

SIPRI is an independent institute for research into problems of peace and conflict, especially those of disarmament and arms regulation. It was established in 1966 to commemorate Sweden's 150 years of unbroken peace.

The Institute is financed by the Swedish Parliament. The staff, the Governing Board and the Scientific Council are international. As a consultative body, the Scientific Council is not responsible for the views expressed in the publications of the Institute.

sipri

Stockholm International Peace Research Institute
Sveavägen 166, S-113 46 Stockholm, Sweden
Cable: Peaceresearch, Stockholm
Telephone: 08-15 09 40

Internationalization to Prevent the Spread of Nuclear Weapons

sipri

Stockholm International Peace Research Institute

Taylor & Francis Ltd
London
1980

First published 1980 by Taylor & Francis Ltd
10-14 Macklin Street, London WC2B 5NF

Distributed in the United States of America by
Crane, Russak & Company, Inc.,
3 East 44th Street, New York, N.Y. 10017
and in Scandinavia by
Almqvist & Wiksell International,
26 Gamla Brogatan,
S-101 20 Stockholm, Sweden

British Library Cataloguing in Publication Data

Stockholm International Peace Research Institute
Internationalization to prevent the spread of nuclear weapons.
1. Atomic power—International cooperation
I. Title
621.48 TK9153

ISBN 0-85066-200-1

Typeset by Georgia Origination, Liverpool
Printed and bound in the United Kingdom by
Taylor & Francis (Printers) Ltd., Rankine Road,
Basingstoke, Hampshire RG24 0PR

Preface

The recent International Nuclear Fuel Cycle Evaluation (INFCE) has concluded that, in a world in which an increasing number of countries are using nuclear energy for peaceful purposes, no technical ways exist to prevent the spread of nuclear weapons. Non-proliferation is a political problem and must, therefore, be solved by political means.

The reasons why countries want nuclear weapons—whether to meet their real or perceived security requirements or for political prestige—must be removed. Resolution of regional conflicts would diminish the stimulus for proliferation, while significant nuclear disarmament measures would de-emphasize the role of nuclear weapons in international politics. Meanwhile, institutional arrangements could minimize the risks of nuclear weapon proliferation. One such arrangement could be the internationalization of the sensitive, that is, the most proliferation-prone, elements of the nuclear fuel cycle.

This book analyses some of the political, economic, technical and legal issues involved in internationalizing the nuclear fuel cycle. It consists of two parts. Part I is SIPRI's evaluation of the advantages, as well as the weaknesses, of the existing proposals for internationalization. Part II contains the papers contributed to the SIPRI symposium on 'internationalization of the nuclear fuel cycle', which was held in October/November 1979. This book is intended to serve as background material for the preparation of the 1980 NPT Review Conference, a crucial event in the field of arms control.

The editors of this book are Frank Barnaby, Director of SIPRI; Jozef Goldblat and Bhupendra Jasani, senior members of the SIPRI research staff, and Macha Levinson and Joseph Rotblat, visiting scholars at SIPRI.

Acknowledgements

The editorial assistance of Barbara Adams, Gillian Stanbridge and Connie Wall is gratefully acknowledged.

SIPRI
December 1979

Contents

Energy units

The SI unit of energy is the *joule (J)*, but energy is often measured in *watt hours (Wh)*, which is the amount of energy generated when a plant operates for one hour at a power of one watt.

$$1 \text{ Wh} = 3\ 600 \text{ J}$$

Larger values are usually expressed in multiples of one thousand, using the following prefixes:

$$\begin{aligned} \text{kilo (k)} &= 10^3 \\ \text{mega (M)} &= 10^6 \\ \text{giga (G)} &= 10^9 \\ \text{tera (T)} &= 10^{12} \\ \text{peta (P)} &= 10^{15} \\ \text{exa (E)} &= 10^{18} \end{aligned}$$

The electrical power of a power plant is given as *MW(e)*, or megawatts of electricity, and *GW(e)*, or gigawatts of electricity. This is about one-third of the thermal power of the reactor.

The total electricity generated is obtained by multiplying the electrical power output by the number of hours a power plant is in operation. For example, if a 1 000 MW(e) plant operated for 6 000 hours in one year, the total electricity generated would be:

$$1\ 000 \text{ MW(e)} \times 6\ 000 \text{ hours} = 6 \text{ million MW(e)h, or } 6 \text{ TW(e)h.}$$

Glossary

Advanced gas-cooled reactor (AGR)	A graphite-moderated, CO_2-cooled thermal reactor with slightly enriched uranium as a fuel.
Back-end (tail end) of the fuel cycle	The part of the fuel cycle following the removal of spent fuel elements from the reactor.
Boiling water reactor (BWR)	A light water reactor in which ordinary water, used both as a moderator and a coolant, is converted to high-pressure steam which flows through the turbine.
Burn-up	A measure of reactor fuel consumption. It is expressed as the amount of energy produced per unit weight of fuel in the reactor.
CANDU	A reactor of Canadian design, which uses natural uranium as fuel and heavy water as moderator and coolant.
Centrifuge method of enrichment	An enrichment process in which lighter isotopes are separated from heavier ones by means of ultra-high-speed centrifuges.
Cladding	The material (zirconium alloy or stainless steel) in which the fuel elements in a reactor are sheathed.
Co-location	Location of different facilities (e.g., reactors, reprocessing plants, fuel fabrication plants) at the same site.
Completing the fuel cycle	Recycling of plutonium and uranium.
Coolant	A substance circulated through a nuclear reactor to remove or transfer heat. Common coolants are light or heavy water, carbon dioxide and liquid sodium.
Cooling pond	A pool of water in which the spent fuel elements are placed immediately after being taken out of the reactor to allow the radioactive fission products to decay.
Co-processing	The reprocessing of spent fuels in which uranium and plutonium are extracted together instead of separately.
Core	The central portion of a nuclear reactor containing the fuel elements and usually the moderator, but not the reflector.

Denaturing	The mixing of a fissile nuclide with an isotopic non-fissile nuclide so as to render the former unsuitable for nuclear weapons.
Enrichment	A process by which the relative abundances of the isotopes of a given element are altered, thus producing a form of the element enriched in one particular isotope.
Fast breeder reactor (FBR)	A reactor that operates with fast neutrons and can produce more fissile material than it consumes.
Fissile material	A material fissionable by neutrons of all energies, especially thermal neutrons: for example, uranium-235 and plutonium-239.
Fission	The splitting of a heavy nucleus into two approximately equal parts (which are nuclei of lighter elements), accompanied by the release of a relatively large amount of energy and generally one or more neutrons. Fission can occur spontaneously, but usually is caused by absorption of neutrons.
Front-end of the fuel cycle	The part of the fuel cycle preceding the placement of fuel elements in the reactor.
Fuel fabrication	The manufacture of fuel elements for use in reactors.
Fuel cycle	The series of steps involved in preparation and disposal of fuel for nuclear power reactors. It includes mining, refining the ore, fabrication of fuel elements, their use in a reactor, chemical processing to recover the fissile material remaining in the spent fuel, re-enrichment of the fuel material, and refabrication into new fuel elements.
Fuel element	A rod, tube, plate, or other mechanical shape or form into which nuclear fuel is fabricated for use in a reactor.
Fuel (nuclear)	Fissile material used or usable to produce energy in a reactor. Also applied to a mixture, such as natural uranium, in which only part of the atoms are fissile, if the mixture can be made to sustain a chain reaction.
Gaseous diffusion	A method of isotopic separation based on the fact that gas atoms or molecules with different masses will diffuse through a porous barrier (or membrane) at different rates. The method is used to separate uranium-235 from uranium-238.
Gas-graphite reactor	A nuclear reactor in which a gas is used as the coolant and graphite is used as the moderator.
Graphite	A form of pure carbon used as a moderator in nuclear reactors.
Heavy water	Water in which the ordinary hydrogen is replaced by deuterium.
Heavy water reactor (HWR)	A reactor that uses heavy water as its moderator. Heavy water is an excellent moderator and thus permits the use of natural uranium as a fuel.

Horizontal proliferation	The spread of nuclear weapon capabilities to non-nuclear weapon states.
Irradiated fuel	Nuclear fuel after it has been produced in a reactor.
Isotopes	Nuclides of the same chemical element but different atomic weight, that is with the same number of protons but different numbers of neutrons.
Laser enrichment	An isotope separation technique, in which uranium-235 atoms are selectively excited or ionized by lasers.
Light water	Ordinary water (H_2O), as distinguished from heavy water (D_2O).
Light water reactor (LWR)	A reactor using slightly enriched uranium as fuel and ordinary water both as moderator and coolant.
Load factor	The ratio of energy actually produced to that which would have been produced in a given time had the reactor operated continuously at the rated capacity.
London Club (London Suppliers Club or London Nuclear Suppliers Group)	The group of countries which export nuclear facilities and which meet from time to time to devise guidelines for the supply of such facilities and materials.
Magnox	Magnesium alloy used to can the uranium in some British-built reactor fuel elements.
Magnox reactor	An early version of the AGR, using natural uranium as fuel.
Moderator	A material, such as ordinary water, heavy water, or graphite used in a reactor to slow down fast neutrons to thermal energies.
MOX (mixed-oxide) fuel	Nuclear fuel composed of plutonium and uranium in oxide form.
Natural uranium	Uranium as found in nature, containing 0.7 per cent of U-235, 99.3 per cent of U-239, and a trace of U-234.
Nuclear energy	The energy liberated by a nuclear reaction (fission or fusion) or by radioactive decay.
Nuclear power plant	Any device or assembly that converts nuclear energy into useful power. In such a plant, heat produced by a reactor is used to produce steam to drive a turbine that in turn drives an electricity generator.
Nuclear reactor	A device in which a fission chain reaction can be initiated, maintained, and controlled. Its essential component is a core with fissile fuel. It usually has a moderator, a reflector, shielding, coolant, and control mechanisms.
Nuclear threshold countries (near-nuclear states)	Those states which are now technically and industrially able to develop and manufacture nuclear weapons or other nuclear explosive devices or will be able to do so in the near future.
Nuclear waste	The radioactive products of fission and other nuclear processes in a reactor.

Nuclear weapons	A collective term for atomic (fission) bombs and hydrogen (thermonuclear) bombs. Any weapon based on a nuclear explosive.
Nuclear weapon states (NWS)	Those states which had manufactured and exploded a nuclear weapon before 1 January 1967, i.e. the USA, the USSR, the UK, France and the People's Republic of China.
Once-through cycle	A nuclear fuel cycle in which the spent fuel elements are not reprocessed for the purpose of recovering the fissile materials uranium-235 and plutonium-239.
Plutonium (Pu)	A radioactive, man-made, metallic element with atomic number 94. Its most important isotope is fissile plutonium-239, produced by neutron irradiation of uranium-238. It is used for reactor fuel and in weapons.
Power reactor	A reactor designed to produce useful nuclear power, as distinguished from research reactors.
Pressurized water reactor (PWR)	A light water reactor in which the water serving as moderator and coolant is prevented from boiling by high pressure. It has a secondary circuit to produce steam to drive the turbine.
Prior consent (prior approval)	The permission required from a supplier before a recipient may retransfer certain materials, equipment or technology obtained from that supplier; or before a recipient may reprocess, enrich or otherwise alter supplied nuclear material or material derived from it.
Rad	A unit of radiation dose corresponding to the absorption of 0.01 J per kg of tissue.
Radioactive decay	The gradual decrease in radioactivity of a radioactive substance due to nuclear disintegration, and its transformation into a different element. Also called radioactive disintegration.
Radioactivity	The spontaneous decay or disintegration of an unstable atomic nucleus.
Recycling plutonium	The re-use in fresh fuel elements of plutonium which has been extracted from spent fuel by reprocessing.
Reprocessing	The chemical and mechanical processes by which plutonium and the unused uranium-235 are recovered from spent fuel elements.
Research reactor	A reactor primarily designed to supply neutrons or other ionizing radiation for experimental purposes. It may also be used for training, materials testing and production of radionuclides.
Retransfer	The re-export of materials, facilities, components or technology.
Safeguards	Sets of regulations, procedures and equipment designed to prevent and detect the diversion of nuclear weapons for unauthorized purposes.

Spent fuel element	Fuel element that has been removed from a reactor after several years of generating power.
Spiking	Methods of making plutonium less suitable for a nuclear explosive, or less accessible to diverters, by mixing it with other radioactive substances.
Tail assay	The percentage of uranium-235 left in the depleted uranium after passing through the enrichment plant.
Thermal reactor	A reactor in which thermal neutrons are used to produce fission.
Thorium (Th)	A natural radioactive element with atomic number 90. The isotope thorium-232 can be transmuted to fissile uranium-233 by neutron irradiation.
Tonne, ton	1 tonne = 1 000 kg; 1 ton = 2 240 lb = 1 016 kg. The difference between ton and tonne is often neglected.
Trigger list	A list of materials, facilities, components and technology drawn up by the London Club, the transfer of which would require the application of IAEA safeguards.
Uranium	See Natural uranium.
Vertical proliferation	The qualitative and quantitative capacities of existing nuclear weapon states.
Weapons-grade material	A material with a sufficiently high concentration of the nuclides uranium-233, uranium-235 or plutonium-239, to make it suitable for a nuclear weapon.
Yellowcake	A uranium compound consisting mainly of U_3O_8.

Abstracts of papers

Paper 1. Considerations on the technical outcome of INFCE

U. FARINELLI

INFCE achieved significant, although not spectacular, results. Most technical conclusions were unanimously agreed upon and supply a data base on which future decisions will rely. It is now clear, however, that very different options can be chosen starting from the same data base according to local conditions and national priorities. No really new strategy has emerged from INFCE that would be particularly desirable from the point of view of proliferation resistance and of resource utilization, at least within the time range available to solve these problems. Worthwhile non-proliferation measures are likely to be of an institutional nature, although technical interventions can enhance their effectiveness.

Paper 2. Background data relating to the management of nuclear fuel cycle materials and plants

J. ROTBLAT

In a discussion on internationalization of the sensitive parts of the fuel cycle it is useful to start with a knowledge of the quantities of material involved, and the plant and storage capacities needed. Using INFCE's projections of nuclear power capacity up to the year 2025, along with general data on reactor operation plus some plausible assumptions, the author has calculated several global projections relating to the production of plutonium, the production of spent fuel elements and uranium utilization. With the once-through cycle the uranium reserves may be exhausted by the year 2015, and the total amount of accumulated plutonium could then be about 7 000 tonnes and the total weight of the spent fuel elements about 0.8 million tonnes. Depletion of the uranium reserves would necessitate the introduction of the fast breeder. If the projected nuclear energy production of 4 300 GW(e) for the year 2025 were to be met entirely by fast breeders, the amount of plutonium that would have to be processed annually would by about 10 000 tonnes. The implications for the internationalization of reprocessing and enrichment plants are briefly discussed, and deep concern is expressed about the proliferation problems of a fast breeder régime.

Paper 3. Nuclear fuel cycle internationalization: the uncertain political context

G.I. ROCHLIN

If present trends continue, proposals for international management or control of plutonium and sensitive technologies are certain to be on the international agenda by the time of the second NPT Review Conference in the summer of 1980. There has been little prior discussion as to whether such arrangements are intended primarily to limit exports or legitimate them, supplant national authority or support it, promote equity of treatment among states party to the Treaty or reduce it. Given the fragmentation of present policies on plutonium use, fast breeder reactors, development and trade, there is unlikely to be any general consensus on political goals. New institutional solutions are therefore as likely to be a source of new conflict as a remedy for old ones, particularly if they formally acknowledge the emergence of a condominium of advanced technological states.

Paper 4. An international plutonium policy

A.R.W. WILSON

Existing arrangements appear inadequate to contain proliferation in a situation where non-nuclear weapon states have control over plutonium and reprocessing plants. However, continued development of thermal plutonium recycle and fast breeders is necessary as an insurance against the possibility that uranium supply may be inadequate to meet the demands in the early part of the next century. The plutonium dilemma is to find a way of allowing the technical development effort to proceed without prejudicing the willingness of states to develop and participate in new non-proliferation arrangements. It is suggested that a realistic international plutonium policy would seek to confine plutonium recycle and fast breeder technical development to a few large states, while promoting international efforts to develop arrangements within which nuclear power programmes utilizing plutonium recycle and fast breeder reactors could proceed without posing unacceptable proliferation risks. The international efforts would be directed to strengthening the political fabric of the existing non-proliferation régime, to building confidence in the long-term security of nuclear supply, to seeking agreement on technical and institutional arrangements for plutonium activities, to planning and providing multinationally controlled reprocessing facilities and to replacing bilateral export controls by obligations embodied in multilaterally agreed instruments.

Paper 5. Internationalization and international control of nuclear facilities

B. SANDERS

In trying to clarify some concepts in the present discussions, the paper points out that nuclear proliferation not only raises the risk of nuclear war, but is undesirable because it upsets the international balance. Both the possession by a state of nuclear weapons as well as the mere fact that it has certain sensitive installations that may give it the capacity to make such weapons are relevant. The non-proliferation régime is essential to dispel suspicions that such installations are being abused. The first proposals for an international non-proliferation régime, made in 1946, foresaw the need for physical control of certain nuclear facilities. The IAEA, established 10 years later, does not have the far-reaching mandate of the body originally proposed, but has the statutory mandate to assume some measure of control. Its principal non-proliferation function is the application of safeguards designed to detect non-compliance with an undertaking not to use specified nuclear items for the manufacture of nuclear explosives. Efforts are being made to strengthen the non-proliferation régime by extending safeguards coverage, finding less proliferation-prone fuel cycles (INFCE) and introducing control measures to prevent possible abuse. The ultimate safeguard lies in the effective multilateral control of operations involving special fissionable material. It is this control which should be central to a discussion of the purposes of 'internationalization of the fuel cycle.'

Paper 6. A new international consensus in the field of nuclear energy for peaceful purposes

A. J. MEERBURG

Elements are enumerated for a new international consensus in the field of nuclear energy for peaceful purposes. Such a new consensus seems a necessary condition for an effective long-term non-proliferation policy. Measures must be taken both to increase technically the proliferation-resistance of the nuclear fuel cycle and to make the misuse of peaceful nuclear energy politically more difficult by the establishment of institutional barriers. Except in fast breeder research, the large-scale use of plutonium should preferably be avoided for the coming decades. Thermal recycle should be halted while the amount of reprocessing should be adapted to needs. An international plutonium storage régime needs to be established to pave the way for universal nuclear export conditions, agreed to by both suppliers and recipients. Such a set of requirements should not include the demand for prior consent with respect to reprocessing. Sensitive parts of the fuel cycle, such as enrichment and reprocessing, could be multilateralized; countries would need to accept more intrusive safeguards. Assured fuel supply systems must be further developed.

Paper 7. Some factors affecting prospects for internationalization of the nuclear fuel cycle

W.H. DONNELLY

The paper identifies and discusses several factors likely to affect support for internationalization of the nuclear fuel cycle by the United States and other countries. After examining notable changes in US non-proliferation policies during 1976, 1977 and 1978, attention is called to INFCE and to the 1980 NPT Review Conference as notable factors that can affect internationalization. The analysis then considers several other factors, including the outlook for the world's economies and energy suppliers, the future of nuclear power and the current state of US nuclear influence. The author concludes that the promise of internationalization provides reason to proceed, but cautions that many factors can influence the ultimate outcome. It is suggested that these need attention to avoid surprises and the temptation to assume that the expected non-proliferation benefits of internationalization are sufficient in themselves to assure its success.

Paper 8. International plutonium policies: a non-proliferation framework

D.W. CAMPBELL and M.J. MOHER

This paper reviews those elements of the current non-proliferation régime which bear on reprocessing and plutonium use. A non-proliferation framework is suggested which would recognize that some reprocessing and plutonium use will take place in the years ahead. The paper concludes that an international consensus on effective and common criteria for reprocessing and plutonium use is highly desirable and might be based on: (*a*) a recognition that reprocessing and the resulting separated plutonium pose a proliferation risk which merits specific measures to minimize that risk; (*b*) while respecting the sovereignty of nations and their development needs, a recognition that reprocessing should take place only when, where and to the extent justified by national or multinational programmes; (*c*) a recognition that criteria should be developed by the international community concerning the 'where', 'why', 'how much' and 'how' of reprocessing and plutonium use; (*d*) a commitment by the countries concerned to ensure that international safeguards can be effectively applied to reprocessing facilities, plutonium stores and other facilities where plutonium is used, and (*e*) a commitment to apply adequate physical protection measures. Such a consensus would provide a framework within which the internationalization of certain fuel cycle activities could take place.

Paper 9. Export of nuclear materials

R.W. FOX

The nuclear power industry has yet to find adequate public acceptance because of a lack of confidence relating to: (*a*) peaceful use; (*b*) safety of operation; (*c*) disposal of wastes, and (*d*) physical protection of plants and materials. Consumers and suppliers should have a vested interest in ensuring use for peaceful purposes only. This can be done only through internationally agreed mechanisms. Existing agreements have not created a sufficient degree of confidence that there will be no diversion of weapons-usable material, and a programme of 'supplemental' measures is called for. This could include: (*a*) a scheme for the international control of excess plutonium; (*b*) a scheme for the international control of excess, highly enriched uranium; (*c*) a scheme for the international control of spent fuel; (*d*) joint participation by several countries in sensitive processes, and (*e*) finalization of existing proposals for physical protection. We should aim to move away from a situation in which special restrictions are imposed through bilateral agreements. Trade agreements should be as free as possible from arbitrary termination or suspension. Countries should be given the technical assistance they need. However, nuclear technology cannot be delivered in lumps and needs a strong and advanced base in the purchaser country.

Paper 10. The role of institutional measures in strengthening the non-proliferation régime

S. LODGAARD

The paper reviews present regulation of the international nuclear market. It emphasizes the need to reconsider the policy of unilaterally imposed restrictions and to prepare the ground for negotiation of mutually accepted restraints by suppliers and importers. The importance of extending the coverage of the non-proliferation régime is emphasized, and an international arrangement to be substituted for bilateral reprocessing controls is suggested. Possible elements for such an arrangement are: (*a*) a scheme for international plutonium storage, co-location of fuel cycle facilities; (*b*) prohibition of the release of plutonium in pure form; (*c*) restrictions on the number of reprocessing plants, and (*d*) improved safeguards. The policies of technology denial are also discussed, emphasizing that embargoes may well prove counter-productive in the long run and indicating that a softening of embargoes may be thought of in combination with better export controls to reduce the grey market, longer trigger lists and stepped-up sanctions against countries violating the non-proliferation obligations. Finally, the need to clarify the functions of international safeguards is observed.

Paper 11. Energy independence via nuclear power with minimized weapon-proliferation risks

K. HANNERZ

Internationalization of fuel cycle facilities is one of the best proposals for strengthening the non-proliferation régime while realizing nuclear power benefits for a great number of nations. However, unless it is borne in mind that the strongest motivation for an extensive commitment to nuclear power is energy independence, efforts could be counter-productive. For example, if enrichment technology were to be placed permanently beyond national control, nations seeking independence would tend to opt for HWRs, which present greater proliferation problems than LWRs. Thus, instead of indiscriminate insistence on internationalization of all fuel cycle facilities, the emphasis should be on preventing the establishment of further national reprocessing plants, and the development of a proliferation-resistant enrichment technology should be encouraged.

Paper 12. A nuclear fuel cycle pool or bank?

M. OSREDKAR

The idea of a nuclear fuel bank or pool has kept recurring. The IAEA Statute provided an institutional framework for the Agency to operate as a fuel bank, but the relevant provisions have never been made use of. Yugoslavia, at the Twentieth General Conference of the IAEA, proposed a fuel cycle pool, to which countries could contribute according to their resources. The Nuclear Non-Proliferation Act (NNPA) of 1978 authorized steps for the eventual establishment of an international authority for ensuring fuel supply. Support for the NNPA idea of a fuel bank ensued from the INFCE discussions, but only in relation to countries with small nuclear power programmes, since developed countries have other means of fuel assurance. Motives for the creation of a bank differ. The bank proposed in the NNPA is motivated by non-proliferation considerations, while the fuel cycle pool proposed by Yugoslavia is motivated by a desire for the promotion of peaceful nuclear energy in accordance with Article IV of the NPT. The fuel cycle pool, with the aim of assuring the fuel cycle supplies of developing countries, has the advantage over a bank in that it would stimulate the active participation of its members, while at the same time widening the base of responsibility for non-proliferation.

Paper 13. An international fuel bank

D.L. SIAZON, Jr.

There is a basic compromise among the different categories of states party to the NPT. The undertakings of the different parties to the Treaty are governmental commitments, but these assurances fall short of the Treaty requirements, and states have to rely on commercial markets alone for assurances of nuclear supply. In order to remedy this situation and to re-establish the credibility of the NPT nuclear-supply assurances, an international nuclear fuel bank is proposed. This fuel bank would, in addition to ensuring reliability of nuclear supply, also serve as a mechanism for providing developing countries with the special consideration they are entitled to under Article IV.2 of the Treaty. Various elements are suggested for inclusion in the proposed international nuclear fuel bank.

Paper 14. International plutonium storage

M.L. JAMES

It is desirable to find a way of controlling the use of plutonium so that the legitimate aspirations of states to develop fuel cycles using it can be realized, while offering reassurance against a further spread of plutonium-based weapons. International control under a scheme for international plutonium storage (IPS) would be a measure added to safeguards at a stage that is particularly vulnerable to diversion. Detailed proposals for an IPS scheme are now being prepared by an IAEA Expert Group. In this paper the author gives a personal account of the proceedings. Among the Expert Group's proposals is that all plutonium separated in states party to an IPS scheme should be registered with an IPS controlling body. Plutonium excess to states' immediate requirements for specified and safeguarded use in reactors or research would then be placed in internationally controlled storage at reprocessing and fuel fabrication plants. The operator of the co-located plant would be a joint keyholder of the storage vault with international officers stationed at the site. Plutonium would be released with the authorization of the controlling body or delegated officers, provided that the material was to be covered by IAEA safeguards and a checking procedure showed that the plutonium requested would be absorbed rapidly by the proposed use. The state would be under an obligation to return material, not taken into the authorized use, to international storage. The end use would be verified.

Paper 15. Institutional solutions to the proliferation risks of plutonium

J. LIND

Plutonium, a by-product of all energy production based on nuclear fission of uranium-235, has caused special concern from a nuclear weapon proliferation point of view. Technical and especially institutional measures, in addition to existing safeguards and other arrangements, are called for to dispel such concern. The political objectives of such measures would be to enhance confidence that the use of plutonium for other than peaceful, non-explosive purposes will not take place and that programmes for the peaceful use of nuclear energy can be conducted in a safe, economical and predictable manner. A step-by-step approach to achieve these objectives is discussed. The measures indicated include (*a*) agreement on uses and other management practices for plutonium; (*b*) end-use statements; (*c*) international verification of end-use, and (*d*) a full International Plutonium Management Régime.

Paper 16. International storage of spent reactor fuel elements

B. GUSTAFSSON

Projections for the accumulated amounts of LWR spent fuel show there is a need for more storage capacity, and research is under way to improve the old storage technologies and develop new ones. At present, wet-storage facilities are the most common, but several types of dry-storage facility are under development; these are discussed and the advantages and disadvantages of dry storage outlined. Until now the most common approach has been to store spent fuel in at-reactor storage pools, but more away-from-reactor (AFR) facilities are being planned to cope with the spent fuel storage problem. The USA is considering various alternatives in order to increase storage capacity, but details regarding additional AFR capacity have not been settled. FR Germany is planning an AFR facility at Aahaus. France is constructing more AFR storage facilities to be integrated with existing and new reprocessing facilities. Sweden is planning an LWR-AFR facility to be located at Simpevarp; a special ship is being planned for sea transport of spent fuel.

Paper 17. Spent fuel storage

G.I. ROCHLIN

Spent power-reactor fuel contains both the highly radioactive wastes from operation and the plutonium produced. Because of the plutonium, states are generally unwilling to dispose of spent fuel as waste. Interim or long-term storage is therefore being considered as an alternative to immediate reprocessing and separation of the plutonium. Technology for spent fuel storage is relatively straightforward, and national debate over whether and how it should be stored revolves on the political and environmental desirability of the alternatives. Internationalization is being discussed as an element of current non-proliferation strategies. However, it is not clear whether any major effort is required other than a series of bilateral or multilateral arrangements to deter national seizure or diversion. In fact, waste management considerations may dominate such arrangements as are arrived at.

Paper 18. Regional planning of the nuclear fuel cycle: the issues and prospects

B.W. LEE

A brief review of the historical development of multinational participation in the Regional Fuel Cycle Centre planning is presented. The most comprehensive study ever undertaken was made by the IAEA as a Regional Fuel Cycle Centre Study Project. The results of the study were reported at the Salzburg Conference in May 1977. There have been many novel concepts proposed in support of a regional approach. The desirability of multinational participation in nuclear fuel cycle planning is evident in view of non-proliferation, health, safety and environmental considerations as well as technical and economic aspects. However, because of the inherently diverging objectives among the potential participants, there are a number of basic issues, created by the varying degrees of commitment to nuclear power and subsequent fuel cycle requirements, to be resolved. The question of the institutional and legal framework of such a centre and the most difficult problem of socio-political issues that exist in the region must also be considered. The practicability of multinational participation concepts would very much depend on the new international order of nuclear co-operation based on the outcome of INFCE and the second Review Conference of the NPT to discuss the implementation of Article IV as well as future constraints. The very foundations of such co-operation should be based on a spirit of mutual trust and confidence. Without such mutual confidence, any unilateral implementation of the requirements for bilateral co-operation is doomed to failure.

Paper 19. Multinational arrangements for enrichment and reprocessing

I. SMART

Multinational arrangements for 'sensitive' fuel cycle processes (enrichment and reprocessing) are distinct from, and more difficult than, arrangements to control 'sensitive' materials. Because they touch economic and industrial interests so closely, they must offer economic and industrial benefits commensurate with their non-proliferation effects. In the non-proliferation context, they can reinforce international safeguards as a means both of deterring national abrogation of commitments and of demonstrating exclusively civil intentions. In the industrial context, as existing process consortia demonstrate, they can offer a variety of advantages over independent national activity. Their non-proliferation and industrial benefits must, however, be balanced and related by reference to consistent criteria of restraint, viability, symmetry and parsimony. At the same time, the vexed issue of technology transfer must be faced, bearing in mind that the choice may be between deferred but uncontrolled dissemination of 'sensitive' technology and deliberate transfers under multinational regulation. In addition, process arrangements must be seen to meet the needs of an actual market, and, in their institutional forms, to offer scope for industrially realistic choice by participants and others. Unless they achieve such flexibility, national enrichment and reprocessing programmes will inevitably multiply.

Paper 20. Sanctions as an aspect of international nuclear fuel cycles

P. SZASZ

The possibility of imposing sanctions, while perhaps not absolutely essential, constitutes a most important feature of any international control system. However, the implementation of effective sanctions is a most difficult international enterprise. A nuclear control system culminating in sanctions comprises several elements: (*a*) the assumption by or the imposition on states of obligations not to proliferate and to co-operate with the control system, and safeguards to check on compliance with these obligations; (*b*) decision-making organs to determine whether violations have occurred and what the reactions to these violations should be; (*c*) sanctions, and (*d*) procedures and formulae for distributing the resulting costs. Several of these elements become simpler or more effective if the control is exercised in respect of the national part of a partially internationalized fuel cycle. In particular, the effectiveness of nuclear sanctions, against which a nuclearly autarkic state may be able to shield itself, can be greatly enhanced if part of the fuel cycle the state depends on is internationalized. Also the cost of imposing sanctions, which may be a substantial deterrent to the international community when faced with a nuclearly self-sufficient state, will be much reduced in respect of states dependent on international facilities.

Paper 21. Internationalizing the nuclear fuel cycle: the potential role of international organizations

K.H. LARSON

The paper explores the potential role of international organizations in the development of institutional arrangements for the nuclear fuel cycle. It reviews the historical development of such arrangements and outlines the present status of discussions for the development of new arrangements. It suggests that international organizations, such as the International Atomic Energy Agency, can play an important role if their expertise and capabilities are properly used. A possible role for the IAEA is foreseen in the evolution of a new international consensus in the nuclear energy régime through the provision of a forum for discussions by states and through the establishment of arrangements within the Agency's programme.

PART I

Internationalization to prevent the spread of nuclear weapons

I. Introduction

The choice of plutonium as nuclear fuel offers the tantalizing prospect of energy independence, with early liberation from oil and even from uranium producers. As plutonium is a product of all currently operating nuclear reactors, it has the attraction of being potentially available. Thus some states are being led towards decisions which may make it impossible to control the spread of nuclear weapons. They want to preserve the option of recycling plutonium in thermal reactors and of using it in fast breeder reactors, and the disturbing assertion is being made that plutonium can somehow be made a safe fuel. But plutonium is also a fissile material, and only a few kilograms are needed to make a bomb. Its widespread use would necessitate equally widespread and very strict control in order to prevent its transformation into nuclear weapons.

Such control would be extremely difficult to implement. However, a possible way would be through a single international authority under a scheme for internationalization of all the 'sensitive' parts of the nuclear fuel cycle. That is, the international authority would manage and control all sensitive processes and materials, and ideally only non-sensitive processes would be managed nationally. Of course, full internationalization could only come about gradually. But it is important that the foundations of internationalization be laid soon, before the various schemes for completing the nuclear fuel cycle reach unmanageable proportions. Not only would internationalization facilitate current non-proliferation efforts, but it could perhaps play a part in reducing the motivation for a 'plutonium economy'. The less proliferation-prone once-through cycle might be made more attractive with internationalization, as an international authority would be responsible for spent fuel management, and, by means of a fuel bank, it would offer supply assurance. Thus we could perhaps defer, or even avoid, the frightening hazards of a fast breeder régime, which would have the potential to produce annually enough fissile material to make a vast number of Nagasaki-type bombs.

II. *Proliferation dangers in a 'plutonium economy'*

Today's world is only beginning to recognize the complexities of this potential problem. The nearly 200 thermal reactors now operating have so far produced about 150 tonnes of plutonium, but most of this plutonium is still in the reactors or in spent fuel stocks. For use as commercial fuel this plutonium would have to be retrieved from the spent fuel by reprocessing and then made into usable reactor fuel. To date very little reprocessing has taken place, and the capacity of existing plants is small. Less than 1 000 tonnes of spent fuel could be handled annually, which means that about 8 tonnes of separated plutonium could be produced every year. Even considering planned reprocessing capacity, only a small part of the plutonium that will be contained in spent fuel could be extracted. These are already significant amounts in terms of proliferation, but they are only a fraction of the stocks which would circulate if plutonium were to become a widely used fuel.

The ramifications of entering into such a plutonium economy would become clear if plutonium were to be reprocessed on a large scale and recycled in thermal reactors, even without the advent of the fast breeder reactor. A look at the nuclear energy projections for the end of this century should enable one to visualize the quantities of plutonium in question. The completed report of the International Fuel Cycle Evaluation (INFCE)[1], which contains a projection of future nuclear power growth, estimates that by the year 2000 about 1 000 GW(e) of nuclear power will be produced by the world's reactors. This amount of energy would leave behind in the spent fuel elements about 230 tonnes of plutonium annually.

The crucial question is what would happen if countries were to reprocess the spent fuel to recover these 230 tonnes of plutonium and recycle it. The quantity of fissile material significant for weapon purposes is so small compared to the vast quantities of materials that would be reprocessed that no safeguards could detect some diversion. Moreover, huge shipments of plutonium would travel from reactors to reprocessing plants, then on to fuel refabrication plants and back to reactors. Consequently, measures of physical protection for both the plants and the transports would become one of the most critical problems facing international society.

Police measures of draconian proportions would be required, and the police would have to be equipped with sufficiently powerful weapons to repel possible attacks by armed terrorist groups. The task would probably be assigned to specially trained army units in view of the threat to national security. The continued presence of such heavily armed forces to protect a civilian undertaking could seriously threaten the fabric of democratic institutions. Societies dedicated to upholding basic human rights might not be able to withstand the necessary degree of control exercised by these police or armed forces.

---

[1] This two-year conference initiated by the USA brought together over 60 countries to study many of the issues of nuclear energy especially as they relate to nuclear weapons.

The development of the fast breeder would greatly increase these dangers. Even larger amounts of plutonium, about 10 times more than for thermal reactors, would have to be reprocessed for the same amount of electricity generated. If by the year 2025 fast breeders were to supply all the nuclear energy projected, about 10 000 tonnes of plutonium might have to be reprocessed annually. The management of such colossal amounts of plutonium would present overwhelming difficulties. Thus, a decision to undertake large-scale reprocessing for the recycling of plutonium or for the fast breeder is much more than an economic or technical question. It would dispel any hope of containing the proliferation of nuclear weapons with any degree of confidence. Moreover, it would have far-reaching implications for the political organization of society in the 21st century.

III. Proliferation dangers today

The dangers of the uncontrolled spread of plutonium do not belong only to the future. They are already present today in the plans for reprocessing and the acquisition of reprocessing plants. A country with this technology could gain access to fissile material directly usable for making nuclear weapons. Plutonium is, in fact, the substance most used in nuclear weapons with the smallest material requirements. With a reactor to make plutonium and a reprocessing plant to recover it from the spent fuel, a country would be in a position just short of the weapon stage. Only a political decision would be needed to manufacture a nuclear explosive device. Moreover, the search for effective controls over this process has to date been unsuccessful. Assuming the presence of the necessary non-nuclear components, the time between the decision to produce an explosive and its actual production could be compressed into several hours, thus eliminating the possibility of timely detection of the diversion by international safeguards. Of all the steps in the nuclear fuel cycle, the separation of plutonium from the other waste products is the most proliferation-prone. Diversion would be most easily effected during the latter part of the separation process when the plutonium becomes more accessible and thus becomes a potential target for theft by non- state groups as well. The incipient spread of this so-called 'sensitive' technology is already causing alarm, and it is feeding uncertainty about where or when proliferation will occur.

Access to fissile materials through the peaceful nuclear cycle may thus be the key to a weapon capability for many non-nuclear weapon countries, and control over the availability of fissile materials has become necessary for non-proliferation purposes. But the measures being taken to prevent the spread of nuclear weapons cannot cope with this new development. More than 110 countries adhering to the Treaty on the Non-Proliferation of Nuclear

Weapons (NPT) are obligated either not to transfer nuclear weapons or not t develop them. These commitments are controlled by the safeguards of the In ternational Atomic Energy Agency (IAEA) accepted by the non-nucle weapon parties on all their peaceful nuclear activities. However, neither th NPT nor the safeguards can prevent a country with a reprocessing plant fro having access to fissile materials. Unfortunately technology has not yet pro vided a solution for this dilemma. It has been impossible to find proliferation-resistant process for plutonium separation. Thus, the onl alternative would be to restrict or defer the technology in question.

IV. *Technological choices to support non-proliferation*

Restraint in reprocessing may be brought about in fact by natural causes. Th growth of nuclear power has slowed considerably since the optimistic projec tions of the 1960s and even the early 1970s. Rising costs, increasing publi resistance and increasing awareness of the insoluble problems which nuclea energy entails have dampened much of the original enthusiasm. Moreover, i is unlikely that nuclear energy will be able to supply even 10 per cent of worl energy needs by the year 2000. Known and estimated uranium reserves coul easily meet these needs. Thus reprocessing and recycling plutonium in therma power reactors would have no economic justification—a conclusion that ha also emerged from the INFCE study.

Much of the drive for reprocessing has been motivated by the quest fo energy independence based on the assumption that either recycling o plutonium fuel or eventual acquisition of the fast breeder would enable country without uranium reserves to become self-sufficient in electricit generation. There are, however, strong economic arguments against bring ing nuclear energy and plutonium-fuelled reactors into additional countries particularly less-industrialized countries. The very large initial investmen necessary as well as the inability to benefit from the economic advantages o large reactors could make this a liability rather than an asset for th economy. Moreover, it is generally recognized that for the fast breeder t become part of a national nuclear energy programme, a large previously in stalled nuclear infrastructure, involving a minimum of 20 or more therma reactors, would be required. Given the projections for nuclear energy in th world over the next 30 years, this excludes the fast breeder for mos countries in the foreseeable future. Under these circumstances invest ments in reprocessing plants would be wasteful.

Thus the only reason for recycling plutonium would be to service th handful of countries which will eventually have fast breeders. Since thes reactors will not be operational on a commercial scale for several decades, halt to reprocessing will not prejudge any future energy decision states may want to make. It will alleviate a serious proliferation threat and allow time for reflection on the economic value and environmental hazards of the fast breeder. It would now seem sensible, therefore, to defer any reprocessing until there is a proven need for plutonium fuel.

All these reasons which militate against the use of plutonium fuels confirm the conclusion that the adoption of the once-through cycle is the only logical policy. In this cycle the spent fuel elements, when removed from the reactor, are placed in storage, and there is no separation of plutonium from the waste products. The present generation of thermal reactors operating on the once-through cycle without reprocessing could more than adequately meet nuclear energy needs for the next decades. The decision to acquire nuclear weapons is and will probably always be a political one, but the technical choice of the once-through cycle would hinder access to the fissile materials and make the acquisition of such weapons considerably more difficult.

V. Internationalization as a non-proliferation measure

Technological choices will not suffice, however, to eliminate all risks of nuclear weapon proliferation. It must be recognized that both reprocessing and the development of the fast breeder are in fact taking place, and it must realistically be assumed that they will continue even if only to a slight degree. On this limited scale they already could abet weapon developments because of the very small amount of plutonium required to make a nuclear weapon.

Uranium enrichment is another sensitive technology which could supply the raw materials to make atomic bombs. The same technology which can enrich uranium to the 3 per cent level needed for reactor fuels could also serve to make weapons-grade material by further enrichment. The spread of indigenous enrichment technology is thus another potential source of fissile material in non-nuclear weapon countries. No safeguards system devised can give the necessary confidence that diversion is not taking place in either enrichment or reprocessing facilities. Thus, both of these technologies, as they spread, will do so without adequate controls.

Yet the overwhelming majority of countries have recognized the danger that proliferation of nuclear weapons will increase the possibilities of nuclear war. Therefore preventing the intentional or inadvertent abuse of peaceful nuclear activities to produce weapons is an international problem and must be handled at this level. Moreover, there is now general recognition of the need for international institutional measures to minimize the proliferation risks of peaceful nuclear power.

Internationalization of the sensitive components of the nuclear fuel cycle would be an effective response to this need. It would be a means of encasing these technologies and assuring that the most proliferation-prone activities of peaceful nuclear energy production were operated under conditions established by the international community. It would promote mutual confidence among nations by allaying the fears of misuse of peaceful nuclear technology. Specifically, it would be possible to improve controls over the handling of plutonium. If the drive towards the plutonium economy should nevertheless continue, it would do so under the guidance of the world community.

Internationalization could also resolve the apparent contradiction between policies seeking to prevent the spread of nuclear weapons and those seeking to develop peaceful nuclear energy. The NPT contains provisions supporting both objectives, but developments over the years seemed to make their implementation incompatible. The measures taken by some suppliers to strengthen proliferation barriers have been interpreted by the recipients as contrary to the pledges of nuclear co-operation and assistance. The serious rift which has developed along North–South lines now presents a real threat to the NPT's survival. The disagreement is not over non-proliferation goals, but over the extent to which nuclear energy technology needs to be restrained in the name of these goals. The proposal for internationalization would provide all parties with the assurances needed for peaceful nuclear uses, and it would also help to resolve the problems of waste disposal.

VI. Three attributes of an internationalization scheme

Three major concerns would be met by internationalization. Firstly, it is necessary to buttress the NPT in its basic requirement that peaceful energy will not be used for weapon purposes. The NPT can no longer satisfy this requirement, partly because it cannot deal with the potential for weapon proliferation inherent in the expansion of nuclear energy. The imperceptible nuclear weapon capability of a country possessing one of the technologies to make fissile materials is not contrary to the NPT, which prohibits only the acquisition of nuclear weapons or nuclear explosive devices. Yet the suspicions and fears that such a country is about to test or develop a weapon are now as real as the act of acquiring the weapon. This fear is compounded by the fact that some countries with significant nuclear activities (so called 'threshold' countries) have remained outside the Treaty. Thus, the proliferation threshold, the point at which a country is a potential nuclear weapon state, is constantly being lowered and is directly affected by the possibility of access to fissile materials.

The nuclear suppliers became aware of these new proliferation dangers at the time of the explosion of a nuclear device by India in 1974. To a large extent the suppliers had brought the situation on themselves. Over the years they had continued to supply non-parties to the Treaty without requiring comprehensive safeguards as stipulated in the NPT. Some of these non-parties acquired or are now planning to acquire the sensitive facilities which could bring them a nuclear weapon capability.

In attempting to deal with the problem of 'runaway' technology, the suppliers formed the so-called 'London Club' to co-ordinate export policies. Eventually, 15 nuclear suppliers agreed to the Guidelines for Nuclear Transfers which applied stricter conditions to the export of an agreed list of nuclear items and recommended restraint in the transfer of sensitive facilities. The United States took even more stringent steps unilaterally in adopting the Nuclear Non-Proliferation Act of 1978. It required the renegotiation of existing nuclear agreements to include new conditions of full-scope safeguards and prior US consent for retransfer and reprocessing. It also restricted the transfer of sensitive technologies. These measures applied to all recipients and seemed to renege on the nuclear assistance commitments in the NPT. An internationalized nuclear system could restrain national access to sensitive materials and processes in the interests of both suppliers and recipients, without arousing resentment or suspicion of collusion.

The second major concern is that of the non-nuclear weapon states, especially in the Third World, who want assured fuel supplies and services for their growing nuclear energy programmes. These countries are reluctant to surrender the right to any peaceful technology for non-proliferation reasons. They seek the fullest possible development of peaceful nuclear energy and what is explicitly described in the NPT as assistance with regard to materials, equipment and technology. Any technological restrictions on their peaceful programmes, such as the various new conditions attached to their exports by the nuclear suppliers, has been met with indignation. Yet, in most cases, the size of nuclear power programmes in these states does not warrant an independent enrichment or reprocessing facility. What is needed is a guaranteed fuel supply from the countries which have these enrichment capacities. Under appropriate international arrangements, this function would be performed without risk of interruptions, thus eliminating the motivation of acquiring indigenous plants.

The third concern which internationalization could manage would be the disposal of wastes from peaceful nuclear energy activities. Here there is little disagreement among countries that storage of spent fuel after its use in a reactor as well as disposal of radioactive wastes after reprocessing are environmental issues of a global nature. Thus, an international solution to this problem would be welcomed by most countries. It would also be a way to satisfy public opinion that responsible and non-commercial interests were seeking to resolve this dilemma. It might even provide the incentive for some states to support international co-operative measures related to the peaceful uses of nuclear energy.

Wisely conceived international arrangements could thus effectively restore confidence and meet the concerns of both nuclear suppliers and recipients. It would ensure control over the enrichment and reprocessing technologies, and it would provide greater reliability of supplies and services for the non-nuclear weapon states in accordance with their needs. While its main purpose would be to prevent nuclear weapon proliferation, economic, health and safety matters would also be served.

VII. The scope of internationalization

The extent of the international measures to be adopted would depend to some degree on the modes of nuclear energy production. Ideally, the sensitive parts of the fuel cycle should be fully internationalized. This would encompass enrichment facilities, fuel-fabrication plants and storage of spent fuel, as well as reprocessing plants, mixed-oxide fuel-fabrication plants, storage of separated plutonium and waste disposal if any reprocessing were to be carried out. All these materials and processes should be managed and controlled by an international authority. No country would be allowed to undertake any of these activities on a national scale.

Even at today's low rate of nuclear power utilization, such an international system would be useful. It would provide assurances against diversion and help accumulate experience for the operation of a larger international undertaking. The relatively small number of enrichment and reprocessing facilities operating today and the manageable amounts of spent fuel in storage would facilitate the beginning of internationalization. Once these activities have assumed larger proportions, it will become more difficult to bring them under international management.

However, internationalization could not be accomplished in one fell swoop. It would be a process to be initiated at lower levels of international co-operation. The first steps could be specific measures taken independently of each other, but together they could eventually develop into a system of full internationalization. Early co-operative measures could be directed towards handling certain materials or processes of the fuel cycle.

In fact, full internationalization would necessarily have to come about gradually. There are many complex issues, some of which could be resolved only on the basis of practical experience. Technical problems concerning the location of fuel cycle facilities, the types of fuel to be produced, transportation arrangements, size of spent fuel stores and safeguards systems would have to be thought about. Economic questions concerning pricing of fuels and distribution of costs among the members of the international system. would have to be worked out. Legal problems involving those parts of the fuel cycle remaining under national control and those internationalized would also be highly complicated. Inevitably, the political problems would be the thorniest, especially if reprocessing were to increase, focusing on the access to fissile materials and on conditions for their release and use. Thus, internationalization cannot be programmed. It will have to come about in an *ad hoc* fashion, but it is important to set out clear criteria and objectives so that specific steps taken will contribute to building a valid non-proliferation régime.

VIII. The time is ripe for internationalization

There is reason to believe that the time may be ripe for an action-oriented approach to this problem. The recent INFCE conference represented a first attempt to discuss the technical issues of nuclear energy internationally, and although the two-year study was not a negotiation, some agreed facts and common attitudes did emerge. Perhaps the most important result of INFCE is the general agreement that there are no foolproof technical solutions to the problems raised by the proliferation-prone processes of the nuclear fuel cycle. Thus the report recognizes and stresses the need for institutional measures. Another rather surprising outcome of INFCE is the preparation of an agreed technical data base regarding nuclear power, and an unexpected bonus is the distinct appreciation on the part of many INFCE participants of the need for international steps. This emerging consensus should now make it easier to discuss the consequences of the agreed technical facts and seek international action to deal with them.

The United States has already recognized the need for international activity in the field of nuclear energy by adopting the Non-Proliferation Act of 1978, which specifically calls for the establishment of an international nuclear fuel authority and which envisages some form of international arrangement for various aspects of the fuel cycle. Ideas of this nature had already surfaced during the previous US administration, and thus a considerable amount of thinking has already taken place.

Another reflection of the growing interest in international co-operation is the fact that two IAEA groups are considering respectively international plutonium storage and spent fuel management. These functions were written into the statutes of the IAEA in 1957, but remained dormant until recent years. Now more countries are facing nuclear energy issues and realize that many of them need attention on a global level.

The search for international solutions is well under way, within governments as well as in a variety of international forums. The specific problems have been pretty well identified. Public opinion has been fully mobilized and could be receptive to such solutions. This momentum should not be lost. It should now be directed towards the actual working out of agreed institutional arrangements, and countries that have not participated so far must also be brought along.

IX. The road to internationalization

Common practices and codes of conduct

The steps that could be taken at a first level would include national activities which could be co-ordinated on an international scale. Since the biggest problems have arisen from the policies applied to nuclear trade, this is where initial action can be undertaken. Suppliers have in effect established export controls over nuclear materials which seemed contrary to the interests of most recipients. Yet the London Nuclear Suppliers Group, which created some of this resentment, was a distinct step forward towards the co-ordination of export policies, although it did not go far enough in demanding full-scope safeguards. Even greater resentment was aroused by the US Non-Proliferation Act, which was also an attempt to tighten safeguards and to inhibit the spread of sensitive materials and facilities.

These various restrictions on international nuclear trade were motivated by non-proliferation reasons. In this sense they were, in fact, also in the interests of the recipients. The parties to the NPT, who are required to accept safeguards on their entire nuclear programmes, have often protested against the discriminatory implementation of this provision, as it has not been imposed on non-party importers of nuclear items. They would certainly agree that full-scope safeguards should be a minimum requirement for all nuclear imports. However, a serious weakness of the suppliers' initiatives lay in their failure to consult with the recipients. Such measures could perhaps be made more acceptable by prior discussions between the two groups of countries.

Moreover, there is no reason why any country should not be willing to accede to full-scope safeguards unless it is seeking a covert or latent nuclear weapon capability. Such political use of peaceful nuclear energy cannot be tackled directly by international measures. But it might be possible for nuclear energy producers to agree that a country unwilling to accept full-scope safeguards would be regarded as deliberately leaving an option open for acquiring nuclear weapons. This stigma in itself might help to bring some more countries in line to join the NPT. On the other hand, suppliers must maintain uniformity of supply conditions as well and avoid commercial competition leading some to offer more lenient terms than others.

Similarly there should be predictable conditions regarding the requirement for the supplier's prior consent for the reprocessing of nuclear fuels. Suppliers should agree under which circumstances they would or would not allow the retransfer and reprocessing of their supplied materials, and agreement should be reached with the recipients about the necessity for these conditions. Periodic consultations would be necessary as the situation in the nuclear industry changed and more or less reprocessing became prevalent or necessary.

As long as trade agreements remained bilateral, these common practices and criteria could even be formalized into guidelines or codes of conduct, with the participation of suppliers and recipients. Such agreements do not, of course, contain any enforcement instruments and would thus be less satisfactory than international agreements, but they would be a first step and would at least prevent some of the chaos now existing in the international nuclear market because of differing export policies.

Even at this very first stage of international consultations, the need for mutually acceptable restrictions must be established. Recipients must recognize that restrictions on the transfer of reprocessing or enrichment technology have a non-proliferation purpose because the technology cannot be properly safeguarded. At the same time, suppliers must recognize the need to restrain their commercial interests, abide by their pledges of assured fuel supplies and remain firm on safeguards requirements.

Existing proposals for international measures

International Plutonium Storage (IPS)

Discussions have been undertaken at the IAEA for a plutonium storage scheme which would provide for international storage of excess separated plutonium. These discussions reflect a serious attempt on the part of the participating countries to deal with the most sensitive material in the nuclear fuel cycle and to come to grips with the essential question of plutonium uses. There are 24 countries involved in this Expert Group from both nuclear supplier and recipient countries. They represent East and West, North and South, and perhaps most important, parties and non-parties to the NPT. It can thus be considered a prototype of the kind of negotiation which will be necessary to establish an internationalized fuel cycle.

There are, however, some pitfalls or shortcomings in the concept of International Plutonium Storage (IPS) which could usefully be pointed out. In the first place, the very concept assumes that reprocessing will take place, possibly on a large scale, otherwise international storage would not be necessary. The assumption is also that reprocessing will take place nationally, and thus the very activity where diversion is most likely to take place will remain outside international control. This is, of course, not a failing of the Expert Group, for reprocessing is not within its frame of reference. However, agreement on storage might emit a false sense of security that the dangers of plutonium had been overcome, whereas, in fact, there has been no control over the material in use.

The other problem would be to identify the conditions under which plutonium would be stored and under what conditions it would be released from storage. Because of the small amount of this material needed to make a weapon, it is essential that all plutonium not in use in the reactor be in storage unless it is in transport. Thus, the determination of the quantity of plutonium given for storage is a delicate one. If left up to individual governments, there would be suspicions that some materials had been held back.

On the other hand, a determination by an international organization regarding the amount of excess plutonium a state may have available for storage could be considered an encroachment on national prerogatives, and these would not easily be given up by states. Of equal importance would be the conditions established for release of the plutonium. These would have to be very strict, requiring detailed justification by the requesting state and some way of checking whether its request accorded with non-proliferation objectives. Otherwise the international storage area might serve only as a depot for temporarily unused plutonium. Caution must also be taken to ensure that the arrangement for plutonium storage does not provide an incentive to proceed to reprocessing and fast breeder development.

The international community will be following this negotiation with great interest. It is a pioneering discussion dealing with a weapons-usable material on an international level. It will perhaps be a measure of the willingness of states to concede some national sovereignty for the sake of international objectives. If successful, it could lead to further steps towards internationalizing the reprocessing of plutonium as well.

The fuel bank

Considerable talk has taken place about the possibility of establishing an international fuel bank or pool to furnish especially the developing countries with fuel supplies. As presently envisaged, these schemes emphasize the objective of assuring fuel supplies to countries which have received insufficient assistance or have been victims of unilateral changes in export policies. They are not specifically an attempt to internationalize fuel supply services. However, both concepts include some essential features of an international scheme and could lead to further arrangements.

Both bank and pool schemes address themselves to the needs of developing countries and focus attention on inducements which could bring them into an international nuclear system. The emphasis of the pool is rather on the development of resources and technology in the developing world, while the bank is seeking to ensure supplies in case the commercial market fails, whether for reasons beyond or within the control of suppliers. The two concepts are different, but in both cases non-proliferation purposes might be served. The pool looks towards a degree of energy independence in the Third World, but it would presumably prohibit use of its supplies for weapon purposes.

International arrangements for assured fuel supplies would be an essential part of any non-proliferation régime, and it is the price to be paid for technological restrictions and tightened conditions of trade. An important stipulation, however, would be that the fuel bank limit its membership to NPT parties. In this way it would reinforce the implementation of Article IV of that Treaty dealing with co-operation in peaceful nuclear energy, and it would provide an incentive for joining the NPT, which is still the strongest non-proliferation pledge.

Spent fuel management

Spent fuel management is also a subject under active consideration by the IAEA. Because many countries today lack manpower resources and adequate storage capacity and because public opinion is worried about these stores, international measures to deal with these problems would gain support from many countries. In particular, spent fuel storage might be of interest to countries with no suitable geological formations for storage. Present reactor operations make this a current international problem that needs to be examined as quickly as possible. A scheme for spent fuel storage could be associated with an international guarantee of fuel supplies as a service offered to members of a fuel bank.

The management of spent fuel would offer two distinct advantages as a non-proliferation measure. Firstly, it would provide a service necessitated by the generally accepted view that the recycling of plutonium is uneconomical for thermal power reactors, and it would focus attention on the immediate problem posed by an increasing accumulation of spent fuel as a result of deferring such reprocessing, but would also highlight the relatively lesser dangers of the once-through fuel cycle compared to dealing with plutonium stocks after reprocessing. Secondly, it would service those countries which consider reprocessing a way of reducing the amount of radioactive waste with which they have to deal; international arrangements to handle the spent fuel could provide an alternative to this kind of thinking. If, in addition, any compensation could be offered to countries for relinquishing their title to the plutonium contained in the spent fuel, this would provide an important contribution to non-proliferation.

Multinational schemes for the management of processes

The measures discussed so far have concerned international plans for specific nuclear materials which could be the beginning of a larger international scheme. However, for a more direct impact on the prevention of the further spread of nuclear weapons it is the processes which produce enriched uranium or plutonium that should be internationalized.

To establish some degree of international control over these processes, it would be necessary to operate enrichment and reprocessing plants on a multinational or internationalized basis. Multinational would signify the participation of several countries together in one or more parts of the fuel cycle and would assume the existence of more than one such multinational organization. Internationalization, on the other hand, would mean that there was only one international organization managing all sensitive parts of the fuel cycle, to which all countries engaged in nuclear activities could belong.

Under a multinational arrangement, each group of countries would set up their own policy-making and operational boards, somewhat along the lines of existing consortia, such as Urenco or Eurodif, although there would have to be participation by recipient countries as well. Commitments on non-proliferation objectives would be made by participants upon entering into the arrangement. The membership would then assume collective responsibility for these objectives as well as for safeguards agreements with the IAEA. To strengthen this self-enforcing process, it has been suggested that the multinational abide by internationally or perhaps UN-established codes of conduct, which would set out some general principles of behaviour to increase confidence in their commitment to non-proliferation.

Both enrichment and reprocessing plants could lend themselves to these multinational arrangements. So far, not many countries have undertaken these activities. Existing facilities are ample to supply today's demand for enriched uranium, and the need for reprocessing is still very limited. Multinational management of these facilities could prove interesting for both economic and non-proliferation reasons. The services made available, if and when countries needed them, would obviate the need for costly national facilities, while nuclear weapon proliferation might be effectively inhibited.

Enrichment plants

While there is no direct proliferation threat in the uranium enriched to around 3 per cent that fuels the bulk of power reactors in use, the risk is inherent in the technology itself, which could permit further enrichment of uranium to the point at which it becomes weapons-grade. In addition, new enrichment techniques under development will make it easier to raise the degree of enrichment from reactor-grade to weapons-grade fissile material.

Multinational arrangements for enrichment plants could establish some international control over the use of this technology. If the idea of a fuel bank or pool is pursued, there should be no problem of assuring supplies. Thus, there would be no real incentive for the spread of this technology. It is true that the possibility of a technical solution to proliferation through the enrichment process has been proclaimed. The new technique proposed by France, if proved practical, would make the enrichment of uranium to weapons-grade consistency almost impossible. Under these circumstances international institutional steps might not be worthwhile, for countries could signal their commitment to non-proliferation by adopting this technology. However, it is unlikely that all other enrichment technologies would immediately cease to operate in favour of the new one, and, in any case, the introduction of the new method would take a long time to realize. As things stand now, it is still necessary to consider institutional ways of handling this process.

After the fuel bank, a multinational approach to enrichment plants could thus be the next step towards internationalization. The fact that no new plants are needed would imply beginning with existing plants, and groups could be organized consisting of suppliers and recipients. Initially, fuel fabrication could remain under national control as long as most reactors used only low-enriched uranium. If it proved impossible to eliminate the need for highly enriched uranium for research or other reactors then the possibility of diverting this uranium would require urgent measures to bring fuel fabrication under international schemes. Fuel fabrication would in any case require multinational arrangements if reprocessing and plutonium recycle were to become widespread.

Reprocessing
More important, however, would be international measures to deal with any reprocessing which may take place. As some of this activity is already being undertaken, albeit on a small scale, it would be better to take appropriate measures soon, so that any new plants to be constructed would automatically come under international control. At the moment, plants with commercial-scale capabilities exist only in a few advanced countries, and the investment requirements for new plants are very high. Thus, as with enrichment plants, a first step could be made on a multinational basis by bringing the existing plants under some collective form of control to attempt to minimize proliferation risks. If and when the need for such services on a widespread basis is proven, full internationalization would be necessary to strengthen non-proliferation assurances.

Several possibilities could be envisaged which would not require a serious upheaval in current physical arrangements. One approach would be to internationalize the sites of existing reprocessing plants. In this way they could continue to be run on a national basis, but there would be continual international inspection of the materials coming into and going out of the plants to account for the amount of separated plutonium returned to circulation and to guard against diversion of the plants. If all plants were part of such a scheme, no clandestine reprocessing could take place, and there would be transparency of all operations involving plutonium, thus facilitating the application of safeguards. The separated plutonium leaving the plant would also have to be subject to IAEA safeguards with regard to end uses, because this is the critical point at which it could be used either for reactor fuel or for weapons. The plants would have to continue commercial operation to retain the interests of the investors, but governments would have to accept the responsibility for non-proliferation objectives.

Another possibility which has been studied by the IAEA is the establishment of regional fuel cycle centres. Here the essential purpose would be to service a particular area with reprocessing under an agreement committing the participants to non-proliferation of nuclear weapons and full-scope safeguards. The incentives would be those of economies of scale as well as regional concern over health and safety factors. The question of membership in the regional association would, however, be decisive in its effectiveness as a non-proliferation measure. Firstly, it should preferably be limited to NPT parties. Secondly, its membership should include industrialized as well as developing countries with an open-ended arrangement allowing accession of new members when these deemed it appropriate.

Another important condition would be that the member countries should not belong to the same alliance or defence arrangements: this would not provide adequate assurances for non-proliferation and would only risk heightening international tensions, especially if set up along North–South or East–West lines.

Finally, multinationals have been discussed as groupings of countries of different development levels and different geographic locations. This type of arrangement might be the best from a non-proliferation point of view. The participation of suppliers and recipients could assure a higher level of deterrence because the mutual commitment not to divert materials would place a very high political price on a violation. The membership of some non-nuclear countries from the Third World would be essential, especially those countries next in line for indigenous reprocessing plants, for this would eliminate their need to import the technology. Their presence might also ensure a balance between the commercial and security interests of the participating governments. Ultimately, internationalization would have to follow to give any real degree of confidence in the disinterested nature of the undertakings.

Important considerations

There are certain important considerations, however, which must be borne in mind in any multinational design.

Firstly, it would be necessary to guard against the danger of commercial interests prevailing over non-proliferation objectives. A successful multinational operation would have to satisfy the interests of all the participants. For the plant managers, that interest would be economic feasibility, and this would probably be the case even if the plants were subsidized or government-guaranteed. Optimum economic feasibility, however, might require large-scale reprocessing, and thus the existence of such multinational arrangements might serve to promote reprocessing rather than to keep it to minimum levels. Under the circumstances, there would have to be very close government control over policy decisions, especially concerning safeguards and plant capacity.

Secondly, the greater visibility of any operation which involves more than two countries would make clandestine activities politically more risky, but the possibility of the detection of diversion would become more difficult, especially in large plants. Sanctions should be made part of the arrangements; although less enforceable, perhaps, under a multinational system, because alternative suppliers would be available, these could be written into the initial agreement and would serve as a precedent for later full internationalization.

Thirdly, location of the multinational facilities would be a crucial consideration if new plants were to be constructed. The objective would be to reduce as much as possible the risks of plutonium diversion, and this consideration might seem to indicate that the multinational plants could best be located in nuclear supplier states where, in fact, the only large plants now exist, or in countries which are unlikely to develop any nuclear weapon ambitions. Unfortunately, such solutions would be politically unacceptable, and the search for an appropriate site might present a serious obstacle. In addition to these political considerations, important elements to be taken into account include transport requirements, necessary resources, existing infrastructure and the security of plants. This latter point would also make the political and social conditions in the host country a determinant factor.

From a non-proliferation viewpoint, one shortcoming would be that multinational reprocessing facilities might still not suffice to assure against the misuse of the separated plutonium once it left the plant. Commitments not to use the fissile material for weapon purposes would have been made among the members of the multinational, and IAEA safeguards would be applied. But suspicions about the use of the plutonium might nevertheless be aroused by the national activities of a member. While it might seem advisable to include fuel fabrication processes in the multinational as well, if possible with co-location of the plants, the difficulty of making mixed-oxide fuel elements that could be used in a variety of reactors might render this step impractical.

The most complex and most political issue involved in the multinational approach is that of the transfer of technology. Inevitably, the scheme would facilitate such transfers among the countries that were members of the group. The rationale of those who advocate the idea is that the technology will eventually spread in any case and that it is better to establish international control over this development so that it is not misused for military purposes, even at the risk of speeding up the process. Yet it is also possible that economic competition may develop between multinational groups, even if they are of mixed membership, and the competition could spur technological developments not necessarily guided by non-proliferation considerations. Mutual suspicions and fears between countries might simply be transferred to these groups in view of the prevailing mistrust of 'clubs'. Thus, the arrangement would be of value in a non-proliferation context only if it were completely in the open with well-publicized operations and preferably a commitment to further internationalization. Otherwise, the plan could not be recognized by other countries as operating in the interest of the world community.

X. How internationalization could look

The ideal nuclear power tableau for internationalization would consist of thermal reactors operating on a once-through cycle policy with a minimum of reprocessing and the deferral of fast breeder development. Under these conditions, the biggest proliferation problem would be the enrichment facilities. As no new enrichment capacity is needed, internationalization could mean control over existing plants to guard against any excessive production of weapons-grade uranium. Although no satisfactory safeguards have yet been designed for these facilities, the diversion risks under this picture would be reduced as most of the uranium produced would be enriched to around only 3 per cent, the present low-grade level. Highly enriched materials would be necessary only in small quantities for some research and other reactors. The biggest international problems under these conditions would be the storage of spent fuel and its eventual disposal if, in fact, no reprocessing were to take place.

Fuel fabrication would also come under the international authority, which would assume the functions of fuel supplies and guarantee the availability of fuel elements. If any reprocessing were to be undertaken, it would come under the control of the authority as well. The system would require less radical changes in operations than other proposals, especially since the largest enrichment capacity is in the United States, which has already confirmed its support for an international nuclear fuel authority in its legislation.

The internationalization scheme would look completely different if the fast breeder were to be introduced on a commercial scale. It would even be very difficult to decide on how the arrangements should be made. Large amounts of plutonium would be reprocessed, and the reprocessing plants would, for non-proliferation purposes, have to be centralized and kept to a minimum number. Thus, they would have very large capacities, making it virtually impossible to prevent significant diversion. To avoid moving massive loads of separated plutonium, fuel-fabrication installations to manufacture fuel elements for different types of reactors would have to be co-located with reprocessing plants. Under this arrangement, the biggest problem would be the international transport of plutonium. The huge quantities of spent fuel shipped to the plants and equally large shipments of fuel elements returned to breeder sites would require wide-ranging rigorous measures of physical protection.

To implement this plan, all parties would have to surrender their right to national enrichment, reprocessing and fuel-fabrication plants. Diversion possibilities by states outside the system as well as by sub-national groups would, however, remain and would even increase because of the enormous transports going in either direction.

To avoid these transports of radioactive materials, the reprocessing plants could be internationalized on a regional basis, involving a multiplication of the number of such plants. The dispersal of possible diversion points would in that case place a heavy burden on the ability of an international organization, presumably an expanded IAEA, to apply safeguards and ensure physical security. Undoubtedly, internationalization would become increasingly difficult as the plutonium activities expanded. Yet it would be the only alternative to proliferation.

Sanctions and safeguards

At various levels of internationalization, measures to assure compliance with non-proliferation objectives will be essential. Joining the NPT should be a basic condition for membership in most of these schemes, and thus more countries would come under the NPT safeguards system. However, some provision should be made for international action in case of violation.

Multinational or international agreements should include some provision for sanctions. In view of the important proliferation risks involved in the wide use of nuclear power, withdrawal clauses such as that contained in the NPT would not be advisable, especially if commitments for fuel supplies had been undertaken. A country might accumulate stocks and then simply withdraw from the arrangements. Thus, sanctions would have to be negotiated and terms for their application strictly delineated. The most effective sanctions would probably take the form of withholding supplies, and it is likely that only if a substantial part of the fuel cycle were internationalized could such sanctions work.

The concentration of sensitive facilities under either multinational or international control should make it easier to apply safeguards, and it is likely that international collective responsibility can be assumed in safeguards agreements with the IAEA. Some changes of emphasis, however, would occur in the event of large-scale reprocessing because of the vast quantities of plutonium involved. Adequate material accountancy would in practice be impossible to accomplish, and greater weight would have to be given to containment and surveillance techniques.

Safeguards in themselves will not be able to ensure efficient and timely discovery of diversion. Multinational or international arrangements could compensate for this weakness by providing a greater deterrence against diversion. Any violation of the agreement would be a betrayal of the other partners, and political consequences might be severe. For the same reason, the likelihood of collusion would presumably be reduced. It is thus in the political context of violation of international agreements and the threat of sanctions that deterrence and confidence will have to be sought.

Organizational and related issues

It is likely that the IAEA would remain the primary verification body for any international measures taken in relation to the nuclear fuel cycle. Its staff and functions would probably need to be expanded. If any new fuel cycle plants should prove essential, the IAEA should be involved at the design stage. However, new organizations for the operation of multinational or international fuel cycle plants might be necessary.

The best working arrangement would seem to involve the separation of the operational from the verificational functions and their provision by different organizations. Under full internationalization, this authority should by preference be a United Nations agency.

Intricate arrangements would be necessary between the international body and the host country as well as with the IAEA. A special role of the host country would relate to its responsibility for health and safety. It would therefore have a say with regard to the site, design and operation of the facility. With regard to physical security, one could conceive of an international force guarding a multinational plant. But it is unlikely that the host country would agree that enforcement functions on its own territory should be performed *entirely* by an international body. This raises the question of the relationship between the law enforcement authorities of the host country and the international guard force. The possibility of giving the international facility extraterritorial status should be examined, including immunities for the personnel.

Other aspects of internationalization

Some of the problems involved in internationalization would be of a legal nature. The constituent agreements for multinationally owned and internationally controlled fuel cycle enterprises would have to deal, in particular, with such problems as qualifications for initial participation, procedures for adherence, withdrawal or expulsion, the rights and obligations of the parties, the powers and procedures of the decision-making bodies (separation of questions to be decided by governments at the policy-making level from questions to be decided at the management level), capital contributions, location of the facilities, and settlement of disputes.

The political and economic issues may prove to be more difficult. Even under an ideal scheme for internationalization, the degree of control of any international authority over its members would be derived from their continued commitment to its purposes, as evidenced by their willingness to apply sanctions when necessary. The consequences of serious political divisions within the organization would in this case have grave repercussions for international security.

Moreover the economics of an internationalized nuclear system would be complicated. The cost of the operations would be large and the value of the materials in storage facilities very high. The plants could be centrally managed by the international authority which would buy the raw materials, operate the enrichment, reprocessing and fuel-fabrication processes and sell fuel elements to individual countries. Thus, initial funding would require gigantic sums, and this would be a matter presenting complex problems for negotiation. It is to be hoped that the operations would eventually become financially self-supporting. Even an internationalized system would have more chance of success if it could satisfy the commercial interests of its operators and subcontractors. However, the price of supplies and services should preferably be less than or equal to what expenditure on these items would have been without internationalization. Theoretically, economies of scale should make this possible. But it is also possible that the international community will have to pay the difference between commercial and international operation of the sensitive fuel cycle parts if both the non-proliferation aims and peaceful nuclear energy requirements are to be met.

This then is the case for internationalization of the nuclear fuel cycle. International co-operation on this matter is now of great urgency, while the threshold of the plutonium economy has not yet been crossed. How to control future nuclear energy development in the interests of world peace must be on the agenda of coming international gatherings. The discussion should be launched at the second NPT Review Conference in August 1980, for here it would be relevant. Internationalization would essentially extend both the rights and the obligations undertaken by the NPT parties to assure the prevention of future proliferation.

XI. Conclusions

Internationalization of the sensitive parts of the nuclear fuel cycle would be a way to establish some control over the materials and processes of peaceful energy programmes which now threaten to escape safeguards and bring nuclear weapons within easy reach of many additional countries. It is especially aimed at ensuring the operation of enrichment and reprocessing plants in the interests of international security.

A variety of international measures currently under consideration could provide a starting point for the internationalization process. Yet none of these measures in isolation would suffice to prevent the misuse of fissile materials. Multinational arrangements for enrichment and reprocessing plants might offer some assurances regarding security objectives, while providing efficient operations and satisfying economic interests as well. A more effective approach, however, would be to establish one international authority to operate all the sensitive facilities and release fissile materials only in the form of reactor fuel elements.

It is true that internationalization could only block some of the channels to nuclear proliferation. It could not deal with those countries which might seek nuclear weapons for political reasons, nor with countries which might stay outside an international system precisely in order to foster the uncertainty about their nuclear status. Internationalization could primarily relieve increasingly pervasive fears that any country with a peaceful nuclear energy programme could develop nuclear weapons.

The mutual interest of many countries in finding solutions to these problems, clearly manifested during INFCE, must not be allowed to flag. Despite the recent slowdown in nuclear power development, the next 20 years will witness a significant increase in the number of new reactors as well as an increase in the number of countries coming into the nuclear energy business. What is needed now is a concerted effort on the part of the political leaders in nuclear supplier and recipient states to co-operate on these crucial international issues of the future and accept the responsibilities incumbent on users of this highly dangerous energy source.

PART II

Paper 1. Considerations on the technical outcome of INFCE

U. FARINELLI*

CNEN, CNEN-Casaccia, I-00060 Santa Maria di Galeria, Rome, Italy

I. Introduction

The International Nuclear Fuel Cycle Evaluation (INFCE) has achieved very valuable technical results.

Although the two main opposing strategies, the once-through fuel cycle and plutonium re-use in fast breeder reactors, endured through and survived INFCE, the positions of the participants grew considerably closer than they were at INFCE's outset.

Two principal conclusions emerged from INFCE: firstly, that both strategies can be supported on much the same technical data base, and secondly, that there is no realistic third option, that is, a different strategy which is simultaneously less prone to proliferation risks and generally acceptable in terms of resource availability.

II. Uranium needs and availability

Is a uranium-wasteful strategy, such as the once-through operation of thermal reactors, compatible with the availability of uranium, and, if so, for how long? An answer to this question requires the consideration of three factors: nuclear energy growth projections, the availability of uranium, and the specific consumption of uranium during the selected fuel cycle.

The first factor is certainly the least reliable. At present, any prediction of nuclear growth is pure guesswork. The anti-nuclear movement may well prevent any further growth or even reduce the present level of nuclear

* The opinions expressed in this paper are those of the author and do not necessarily reflect the official policy of his professional affiliation.

power. Nevertheless, drastic shortages of fossil fuel in addition to political preoccupations with energy independence could reverse the trend and create accelerated nuclear programmes.

At INFCE, the projections up to the year 2000 were based on national estimates, obtained from a questionnaire with some interpolations and extrapolations. Several recent international studies were used as the basis for determining both the low and high projections for the period between the years 2000 and 2025. For the World Outside 'Centrally Planned Economies' [CPE] Area (WOCA), the low projected nuclear generating capacity is 850 GW(e) for the year 2000 and 1 800 GW(e) for the year 2025; the corresponding high values are 1 200 GW(e) and 3 900 GW(e), respectively. Even with this wide spread, it is obvious that the actual growth could lie outside the selected range.

On the whole, it appears that the values chosen by INFCE are somewhat lower than those appearing in other studies. This discrepancy probably reflects a general trend towards lower figures and the actual delay in the implementation of nuclear programmes rather than a difference in attitude between INFCE and the other groups. Most unfortunately the CPE countries, although participating in INFCE's work, did not supply information on their long-term nuclear projects, which may substantially affect the world scene.

Compared with the prediction of nuclear energy growth, the second factor, that is, uranium availability, was much easier to assess. The studies carried out by the joint Working Party of The Organisation for Economic Co-operation and Development (OECD) and the International Atomic Energy Agency (IAEA), which have been periodically updated since 1965, have been accepted as the most reliable sources of information. The most recent results of these studies for the WOCA indicate 1.75 million tonnes of uranium as "reasonably assured resources" at low cost (up to US $80/kg) and 0.73 million tonnes at medium cost (between US $80 and US $130/kg). Further explorations are considered likely to yield "estimated additional resources" of 1.54 and 0.8 million tonnes of uranium at low and medium cost, respectively. Low-grade sources of uranium can be, and are being, exploited increasingly. For instance, uranium can be obtained as a by-product of phosphoric acid production, and identified phosphate deposits contain not less than 15 million tonnes of uranium.

The major problem, however, may not be in the total amount of uranium resources; rather it may be in the limits to production capabilities. Present projections of uranium-production capabilities from known resources (including phosphates) peak between the years 1990 and 2000 (at about three times the 1978 value of 39 000 tonnes). Thereafter, this production will decline, due to depletion of some deposits and the shift to lower-grade ores in others. In order to maintain a high uranium production, it will be necessary to find additional resources. However, the OECD–IAEA study estimates that even in the best conditions "a major part of the quoted speculative resources may not be discovered and brought into production until after the first quarter of the twenty-first century''. Production may also be severely limited by environmental considerations, public opposition and political decisions.

The INFCE conclusions point out that uranium-production capabilities will depend on the degree of assurance with which nuclear programmes can be predicted many years in advance. It is unlikely that the necessary exploration and development activities will be started, that the large capital investments will be found and that the environmental and political difficulties will be overcome unless reliable long-term planning of nuclear development is carried out.

The comparison of uranium availability and needs also requires knowledge of the specific consumption of uranium (the third factor) based on the various fuel cycle options. Although considerable debate took place at INFCE, the results are affected by only relatively small uncertainties, since they are essentially technical in nature.

If we consider the once-through fuel cycle in present light water reactors (LWRs), which is the option with the highest specific uranium consumption, a 1 000 MW(e) LWR operating on a 70 per cent load factor needs about 142 tonnes of natural uranium per year.

It is estimated that improvements in this fuel cycle (mostly by increasing the average burn-up) could reduce the consumption to about 120 tonnes per year before the year 2000; a further 15 per cent reduction may be possible over longer periods of time. Heavy water reactors (HWRs) use somewhat less uranium: with current reactors and the once-through fuel cycle, the consumption under the same conditions would be about 120 tonnes per year, and an improved cycle (using slightly enriched uranium) could reduce the uranium needs to less than 90 tonnes per year.

How can all these figures be used to assess the compatibility of needs and resources in order to determine the length of time that the once-through fuel cycle can be relied upon and when it will be necessary to introduce plutonium recycling, fast breeders or other solutions?

Unfortunately, there are no clear-cut answers to these questions, even if the starting points are agreed upon. If one considers the low-growth prediction for nuclear energy and is optimistic about the rate at which speculative uranium resources can be discovered and brought into production, there will be no major difficulty with uranium availability (on a world-wide basis) until after the year 2000, especially if one assumes that the least uranium-consuming option within the once-through strategy will spread rapidly. However, if one uses the high curve for nuclear power growth and is cautious or pessimistic about the increase in uranium-production capabilities, one arrives at the opposite conclusion: a dearth of uranium could surface as early as the year 1990; thus something must be done now, considering the long lead times required for new solutions.

Moreover, the problem becomes even more serious if it is brought to the regional or national level rather than world-wide (or WOCA-wide as the case is here). The present barriers to uranium circulation, which make the uranium market anything but a free one, urge alternative solutions to be found even if the global aspects are left aside. In other words, the uncertainties in nuclear energy production, concerning both the nuclear programmes and the institutional frames in which they will be developed, have a major bearing on setting up national fuel cycle solutions that may not be fully justified on economic grounds and that may present higher proliferation risks.

III. Various reactor and fuel cycle options

In the course of INFCE, several calculations have been performed to evaluate the needs for uranium corresponding to low- and high-growth rates and to different reactor types. In general, scenarios for periods up to the year 2000 were 'predictive' in that they were based on the national programmes described in the answers to the questionnaire mentioned in section II above. For the following years (2000 to 2025), the scenarios were for 'binding', or 'illustrative' strategies, in the sense that they explored extreme hypotheses that, without being physically impossible, are very unlikely (e.g., maximum deployment of fast breeders, or full conversion to a thorium cycle). Such cases are useful in assessing the limits of the various strategies and in enhancing the effects of their introduction, which, in reality, would probably be slower and, at least initially, geographically limited.

As expected, these calculations show that massive deployment of fast breeder reactors, and, to a more limited extent, recycle strategies (with thorium or plutonium) in advanced thermal converters, would make even the high-growth projection compatible with the generally accepted levels of uranium-production capabilities.

Apart from these conclusions, and from the neutral wording of most of the final INFCE reports, some very clear impressions on the prospects of various reactors and fuel cycles were registered by the author of this paper and by many of the INFCE participants.

First of all, plutonium recycle in thermal reactors, and particularly in LWRs, is an option that many countries want to keep open but that very few are considering seriously. The former idea that plutonium recycle was economically attractive at current uranium prices is questionable. The difficulties (both technical and political) connected with reprocessing, although certainly capable of being overcome, are hardly justifiable in terms of the limited uranium savings that can result by using standard reactors. However, fabrication of mixed-oxide fuel and its burning in LWRs is a well-proven technology and can be deployed very rapidly. Therefore, the availability of plutonium represents an insurance against possible shortages of enriched fuel; it has an immediate effect on uranium requirements and is, therefore, more of a political bonus than a technical option.

But fast breeder reactors, and in particular sodium-cooled, PuO_2/UO_2-fuelled liquid metal fast breeder reactors (LMFBRs), of the type which is becoming 'standard', are considered to be a very serious possibility or rather the only possibility that can be assured today if nuclear power has to play a major role for a long time. Whatever differences there are among the INFCE participants about the desirability of large-scale deployment of LMFBRs and the time-scale in which such deployment is envisaged, there is general consensus that such an option is technically viable. Moreover, although there is an economic penalty today in using LMFBRs rather than other reactors, this penalty may be or become negligible when compared to other considerations, especially long-term conservation of resources. Probably no nuclear programme in its initial stage should consider LMFBRs, which in terms of both fuel cycle and technology are only justifiable in a mature programme where many thermal reactors are in operation.

HWRs are a fair competitor of LWRs, but there is no indication that they will gain a much larger share of the market than they now have. Apart from the position of countries that are already commited to HWRs, their popularity at INFCE showed a balance between their increased uranium utilization and the suspicion of higher proliferation risks connected with on-load refuelling and heavy-water technology. High-temperature, gas-cooled reactors (HTRs) certainly have a powerful lobby in several countries, and after an initial unfortunate period they have proved to work well and reliably. However, the plans for implementation of HTRs are lagging behind or disappearing. The general tendency is to regard HTRs as long-term options (especially the very high-temperature version) to be considered sometime after LMFBRs as sources of process heat for industrial applications rather than as competitors of present reactor types.

Other reactor types and fuel cycles have come out of INFCE with either a bad record or a suspended judgement. Not only are all advanced concepts far away from industrial development and commercialization (and the penetration times of new types, even when demonstrated, are of the order of decades away), but also due to present conditions no indication exists that the considerable effort necessary to implement them will be started. This applies in particular to the thorium cycle in present thermal reactors: although the feasibility of such a cycle is beyond doubt, there simply does not appear to be enough incentive to start large-scale development and demonstration. In particular the thorium cycles that are most attractive from the point of view of resource conservation are those requiring reprocessing. No one is apparently going to make the financial investment necessary to demonstrate the reprocessing of thorium fuel on a significant scale, such as that which would be required to bring about meaningful indications on economics, reliability and safeguards capability.

The advantages of other concepts, which might require less development effort, have proved to be either very small or to be counterbalanced by potential disadvantages. For example, the spectral-shift concept (using a mixture of heavy water and light water as a moderator and coolant in a pressurized water reactor, and varying the ratio of the two during the reactor cycle to increase uranium utilization) has shown itself to be both very difficult to backfit to existing reactors and hardly worth the effort of developing new reactors to accommodate this system. Another concept, the so-called 'tandem cycle' (using spent fuel elements from an LWR to fuel an HWR) has proved to be practically unfeasible or, at best, unattractive.

Rather advanced concepts, for example, the fusion–fission hybrids, have hardly been considered by INFCE. The reason for this is that a realistic schedule for their development and deployment would bring their possible impact into the age which is far beyond the time horizon considered by INFCE and perhaps beyond the point where proliferation is a concern (in the sense that the proliferation problems will have to be solved by that time by other means). This situation should not discourage research on such concepts, but it would not make sense to consider them in today's decision-making processes regarding proliferation issues.

IV. Off-reactor parts of the fuel cycle

Today uranium enrichment is not a limitation to the development of commercial nuclear power. The current enrichment capacity actually considerably exceeds its needs, due to nuclear programme delays. Enrichment services are spread and diversified, and the low-enriched uranium needed for the operation of LWRs can be obtained from several sources (the European plants of European Diffusion [Eurodif] and Uranium Enrichment Company [Urenco] and sources in the USA and the USSR). In addition to the classical gaseous diffusion method, the centrifuge method can now be considered as commercial. It is to be expected that, should the need arise, new enrichment plants will be built and that the time needed to build them would not substantially exceed the time required to build the reactors that would use their services. The proliferation aspects of such a situation will be mentioned below.

Reprocessing remains as the most controversial point. Basically, however, consensus exists that the fundamental technology for reprocessing is well established. Even if some difficulties have been experienced when dealing with higher burn-up fuel and economics may be less attractive than was hoped, there is no doubt that large-scale commercial reprocessing can be carried out. Methods to reduce inherent proliferation risks connected with reprocessing have been examined; some of these methods are easily applicable, such as co-location (location on the same site of reprocessing and refabrication plants to avoid transportation), storage and transport of plutonium as mixed PuO_2/UO_2 rather than as PuO_2, and perhaps co-conversion (producing mixed oxides from mixed uranium and plutonium solutions, rather than mixing pure oxides). Other methods, although interesting and worth further investigation, have been considered with scepticism by the reprocessing experts, in particular co-processing (a modification of the reprocessing scheme in which plutonium is always mixed with uranium at all points of the plant). Co-processing is seen as a long-term option that still requires research and development. Other measures, such as the use of radiation barriers (e.g., those that could be obtained by leaving some of the fission products with the fuel, or by pre-irradiating the fuel as soon as it is fabricated, or by adding radioactive materials at some point of the process) are viewed with open mistrust by the advocates of reprocessing. Such measures would create environmental problems, economic and resource utilization penalties and would perhaps also make safeguards more difficult.

Waste disposal, although certainly an open problem, is not going to be a deciding factor among different fuel cycles. The once-through cycle, which, if carried out for an indefinite length of time, would involve final disposal of entire fuel elements instead of separate vitrified radioactive waste, would certainly pose more problems than the reprocessing option, which has been more deeply explored. However, it seems likely that these extra difficulties could be overcome in due course, as already indicated by some preliminary studies. In this case, as in a few others, the INFCE members, even when of opposite views, have carefully avoided recording extreme positions that could be used externally against nuclear power *in toto*.

V. Non-proliferation issues

Although proliferation problems constituted the prime force of INFCE, not many technical conclusions concerning them have emerged in the final reports.

As a very general statement, it can be said that all cycle choices present some proliferation problems: that although at least in some parts of the cycle these problems appeared more severe than in other parts of the same cycle or than in the corresponding part of other cycles, it was not possible to quantify these differences, and that, moreover, such an evaluation is extremely dependent on the circumstances.

Safeguards measures can be applied, at least in principle, to provide a reasonable degree of protection against undetected proliferation. Although technical methods to increase the efficiency of safeguards can be developed or improved, the institutional part of safeguards is probably more important than the technical part. Many institutional measures, in addition to safeguards, could alleviate proliferation risks; however, none of them, or no combination of them, can ensure absolute security.

It was evident that in many cases the spread of a technology would represent a much greater risk than the existence of a safeguard plant based on that technology. For instance, enrichment plants built to produce low-enrichment fuel are not easily or rapidly converted for the production of fully enriched fuel, and such a conversion would easily be detected even by the loosest safeguard. However, easy access to centrifuge enrichment technology and components (and in the future this may apply even more to laser enrichment) could be used for setting up small-scale undercover plants.

The situation is somewhat different in the case of reprocessing plants, where the presence of separated plutonium in bulk form (i.e., requiring measurements of concentrations and masses rather than inventories of items, as in the case of fuel elements) does present some safeguards problems. As mentioned above, alternative technical approaches could certainly reduce proliferation risks; however, even though they are complicated and expensive, effective safeguards are, in principle, possible. But, it is more likely that institutional measures can handle the job.

Minimizing the spread of reprocessing technology would be less effective in preventing proliferation than it would be in the case of enrichment, however. For example, reprocessing techniques for low-irradiated uranium, such as would probably be used for weapon production, are relatively simpler than those for commercial plants which process highly irradiated fuel.

Another problem that raised some controversy concerns heavy water. The fact that the plutonium for the Indian bomb was obtained from fuel irradiated in a heavy water reactor makes people very sensitive to this nuclear material. Is it necessary to safeguard heavy water production or even the technology for its production? Apart from the fact that the production of heavy water is a relatively simple (and known) technique, there are other materials that can be used as a moderator in conjunction with natural uranium. Actually, most weapons-grade plutonium has probably been produced with graphite-moderated reactors. Although the graphite used in nuclear reactors must be highly purified (especially as regards its boron content), similar graphite is widely used and marketed for non-nuclear applications, and it seems unrealistic to control its circulation. Although a heavy water reactor is probably simpler to build than a graphite plutonium-production reactor, it is not clear whether this difference justifies special treatment for heavy water.

The continuous refuelling system of CANDU[1] and other types of

[1] CANDU is a reactor of Canadian design, which uses natural uranium as fuel and heavy water as moderator and coolant.

reactors was blamed for being potentially more exposed to diversion. However, safeguarding this type of fuel management, although more expensive, is probably easier than in the case of bulk refuelling, due to the small number of fuel elements involved in each transfer operation.

Somewhat peculiar treatment was allied to research reactors. Practically the only point that was debated in great detail was the possibility of operating these reactors with low-enriched fuel (less than 20 per cent U-235), probably because both France and the USA have large programmes for the development of such fuel. The conclusion was that in most cases this possibility exists or will exist in the near future, but that in a few cases this substitution would strongly penalize the irradiation programmes.

The highly enriched uranium is used in research reactors in small quantities and is relatively easy to safeguard. However, no discussion took place on the improper use of research reactors for plutonium production, or on the experimental reactors, such as fast critical facilities, that by their very nature include very large quantities of pure plutonium or fully enriched uranium.

Finally, no clear-cut conclusion was reached on the question of the degree of proliferation protection provided by the once-through fuel cycle. Qualitatively, it can be said that the irradiated fuel is protected by its radiation for a certain time, but this barrier decreases with time, and in the long run one accumulates more and more plutonium which is less and less protected. Quantitatively, it is difficult to make precise judgements, which would critically depend on a number of other variables: for instance, the reprocessing capabilities that would be available to the potential diverters. On the whole, the once-through fuel cycle has surfaced again as a possible temporary measure that does not make any sense if prolonged beyond a certain point in time.

Paper 2. Background data relating to the management of nuclear fuel cycle materials and plants

J. ROTBLAT

8 Asmara Road, London NW2 3ST, UK

I. Introduction

For a discussion of problems related to the international management of the materials and plants in the sensitive parts of the nuclear fuel cycle, it is useful to start with a knowledge, even if only approximate, of the quantities of materials involved, and the plant and storage capacities needed. Working Group 4 of the International Nuclear Fuel Cycle Evaluation (INFCE) attempted to obtain the relevant information by sending a questionnaire to governments (INFCE, 1980a); however, due to a considerable number of gaps in the replies, the totals are, in most cases, meaningless. Nevertheless, by using the available data to make some plausible assumptions, it is possible to derive approximate figures that apply to the entire world.

II. Generation of nuclear power

By the end of 1979 the nuclear electricity generated in the non-socialist countries totalled 3 200 TWh (*Nuclear Engineering International,* 1980). The average installed nuclear power in the socialist countries was 9.8 per cent of the world's total (IAEA, 1979). Assuming that the load factor (55 per cent) was the same for socialist countries as for non-socialist countries, the total amount of nuclear energy generated up to the end of 1979 in commercial reactors is 3 520 TWh $= 1.3 \times 10^{19}$J. (The world-wide annual energy consumption is at present about 3×10^{20}J.)

As of the end of 1979, the installed nuclear capacity in the world is about 120 GW(e) net. According to the International Atomic Energy Agency (IAEA), reactors under construction plus those planned will increase the capacity to 422 GW(e) by the year 1993 (IAEA), 1979); more reactors are likely to be ordered and operating by that date, but others may be shut down and some existing orders may be cancelled.

INFCE (INFCE, 1980b) has made projections of the nuclear power capacity in the non-socialist countries up to the year 2025; two sets of estimates were made, a 'low case' and a 'high case'. Based on these estimates and on the assumption that the socialist countries will continue to contribute 10 per cent of the total capacity, the projections for the whole world (INFCE figures multiplied by 1.1) are made and are presented in

Table 1. Nuclear power forecasts GW(e)

Year	Low case	High case
1985	270	301
1990	410	508
1995	605	847
2000	935	1 320
2005	1 210	1 815
2010	1 430	2 365
2015	1 595	2 970
2020	1 815	3 685
2025	1 980	4 290

Figure 1. Projection of nuclear power growth

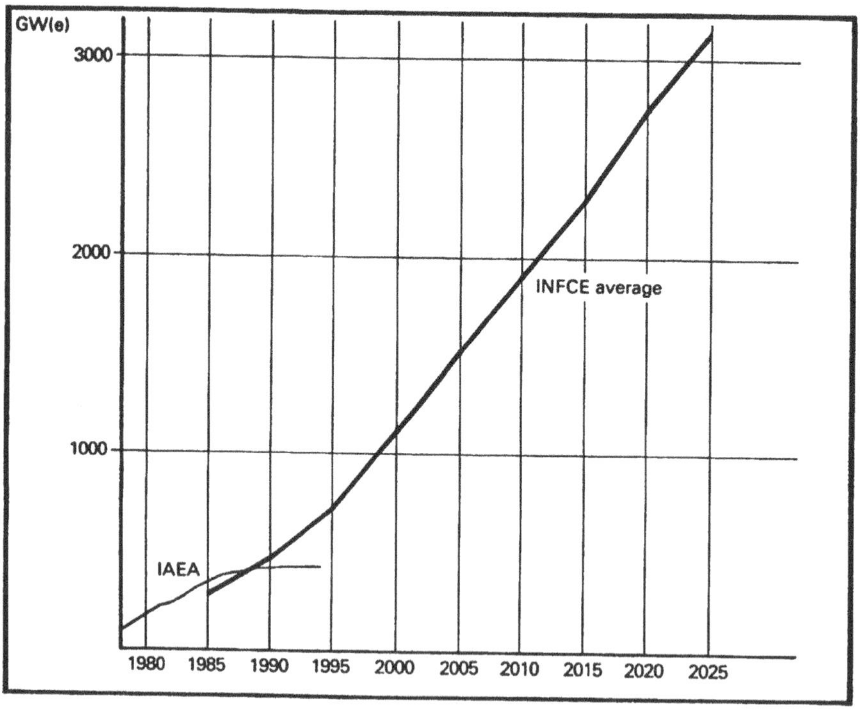

table 1. The mid-values between the low and high cases are shown in figure 1. Starting from the year 1995, the points fit perfectly (and fortuitously) a straight line with a slope of 80 GW(e) per year. Figure 1 also shows the IAEA forecast of nuclear power from reactors already in existence in addition to those under construction and those being planned. It should be noted that until the year 1988 the INFCE projections are lower than the IAEA's; this can be interpreted as a presumption by INFCE that at least 20 per cent of the currently planned reactors will be cancelled.

III. Production of plutonium

Present stocks

The amount of plutonium produced in reactors per unit of energy generated differs somewhat for different reactors. However, since most of the nuclear power (64 per cent) comes from pressurized water reactors (PWRs), we shall not be very wrong if we use the PWR as the typical reactor.

A PWR of 1 GW(e) net power, requiring an annual reload of 27 tonnes of 3 per cent enriched uranium, and working with a load factor of 55 per cent, produces about 200 kg of plutonium annually. (Although only about 70 per cent of reactor plutonium is fissile, here we are primarily concerned with all the plutonium.) On this basis it can be calculated that the total amount of plutonium produced so far in commercial reactors is about 145 tonnes. Some of this plutonium is still in the reactors themselves; if one-third of the core is discharged every year, the above amount will be outside of the reactors after three years.

Research reactors have produced about 3 tonnes of plutonium (IAEA, 1976). If we assume that some 2 tonnes come from other civilian uses (e.g., propulsion of ice-breakers), then the total non-military stock of plutonium is currently about 150 tonnes.

Military stocks

It has been estimated (SIPRI, 1979) that the strategic stockpiles of the nuclear weapon states contain about 20 000 warheads. A similar number has been suggested for tactical weapons. Assuming 4 kg of fissile plutonium per warhead (probably an overestimate), the total military stock is 160 tonnes of fissile plutonium, corresponding to 230 tonnes of reactor-type plutonium. This is larger than the present civilian stock, but with the expected growth of the latter, the two will become equal in the year 1984.

By the turn of the century, military plutonium will be a small fraction of the plutonium produced from civilian sources, unless nuclear armament considerably increases.

Table 2. Production of plutonium (cumulative, tonnes)

Year	Low case	High case
1983	202	202
1988	515	515
1993	969	969
1998	1 552	1 748
2003	2 438	2 994
2008	3 671	4 797
2013	5 190	7 200
2018	6 929	10 268
2023	8 890	14 095
2028	11 072	18 680

Figure 2. Accumulation of plutonium and spent fuels

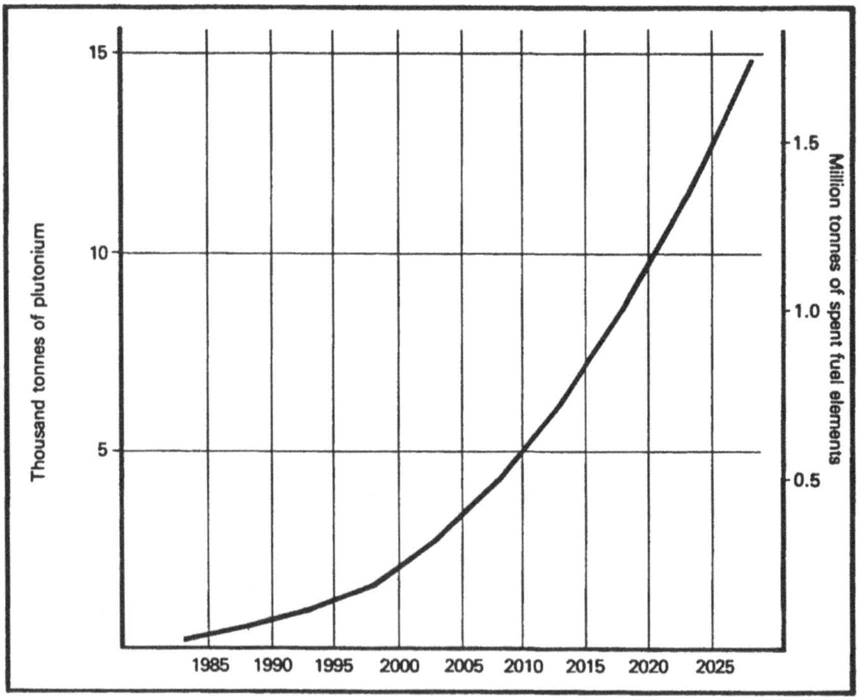

Future stocks

The load factor of PWRs has not increased significantly over the years, and it is unlikely to rise much in the future in view of the many problems experienced with these reactors. But assuming that there will be an increase from the present 57 per cent to 70 per cent for the average future operation, the amount of plutonium produced each year from a 1 GW(e) reactor will then be about 230 kg.

Using the IAEA projections up to the year 1990 and the INFCE projections for subsequent years, and assuming that all nuclear power comes from thermal reactors without recycling, we can then calculate the cumulative amounts up to the year 2028. These figures, which take into account the three-year time lag in the discharge of spent fuel, are presented in table 2 for the low and high projections and in figure 2 for the average projection.

By the year 2028 the total plutonium stock could be between 11 000 and 19 000 tonnes. However, these figures are unlikely to be reached, because of the exhaustion of uranium (see section V); these figures would also be affected if plutonium were to be recycled and commercial fast breeders were to come into being (see section VIII).

IV. Spent fuel elements

Under the same operating conditions, the weight of spent fuel elements is proportional to the weight of the plutonium produced in them. With an annual discharge of 27 tonnes from the typical reactor described above and with plutonium production of 230 kg, the weight of the fuel elements is 117 times greater than that of plutonium. With the once-through cycle, the total weights of spent fuel elements that would accumulate according to the different nuclear power projections can be calculated by multiplying by 117 the corresponding figures for plutonium in table 2 and figure 2 (scale on right).

Thus, by the year 2028, if all the energy produced came from thermal reactors on the once-through cycle, the total weight of the spent fuel elements could be between 1.3 and 2.2 million tonnes, but for the reasons given at the end of section III, the actual weight is likely to be less than 1 million tonnes. The finding of storage space for such amounts would thus be one of the most difficult problems for the once-through régime. However, from the proliferation risk point of view, the fuel elements could be left in national custody for several decades if a generally agreed upon policy of no reprocessing existed.

V. Uranium utilization

For reactors using 3 per cent enriched uranium, produced at a tail assay of 0.2 per cent, the weight of natural uranium metal is 5.4 times the weight of the fuel elements. Therefore, the annual fuel usage of a 1 GW(e) reactor is 146 tonnes of uranium metal. The cumulative uranium requirements can thus be obtained by multiplying the plutonium production by 635 (but

Table 3. Uranium metal utilization (cumulative, thousands tonnes)

Year	Low case	High case
1980	138	138
1985	356	356
1990	672	672
1995	1 043	1 167
2000	1 605	1 958
2005	2 388	3 102
2010	3 351	4 628
2015	4 455	6 575
2020	5 700	9 004
2025	7 085	11 915

Figure 3. Uranium metal used up in reactors

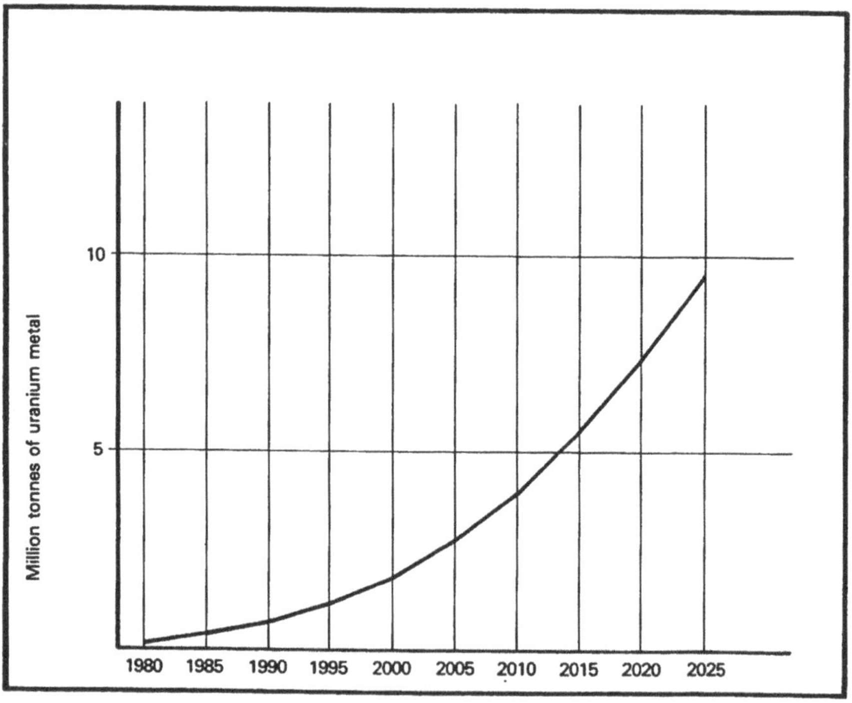

without the three-year time lag). The values thus obtained are presented in table 3 and figure 3 (with the present tail assay of 0.25 per cent, the multiplying factor is 700).

With the once-through cycle the limit to nuclear power expansion is imposed by the availability of uranium ore. In the non-socialist countries the estimated uranium resources are about 5 million tonnes. Allowing a 10 per cent contribution from the socialist countries (probably a gross underestimate), all of this uranium will be exhausted by the year 2015, plus or minus three years. However, it is likely that further exploration will increase the resources. Indeed it has been suggested (IAEA, 1978) that speculative resources may add a further 7 to 15 million tonnes. Should these resources materialize, the thermal reactor régime, on the basis of the average projection, could extend until somewhere between the years 2035 and 2043.

Should the uranium reserves be exhausted in the year 2015, the amount of accumulated plutonium would be about 7 000 tonnes, and the weight of spent fuel elements would total about 0.8 million tonnes.

VI. Reprocessing capacity

The present reprocessing capacity of oxides is very small; less than 1 000 tonnes of spent fuel per year. Plants being built will increase the capacity to about 7 000 tonnes per year. This is a tiny fraction of the future annual increments of spent fuels. Should the decision be made, therefore, to proceed with reprocessing, there is still time to prepare for the international management of practically all reprocessing requirements.

VII. Enrichment

The situation is quite the reverse with regard to enrichment. Currently operating enrichment plants can handle about 50 000 tonnes of uranium metal annually. With the plants scheduled for completion by the mid-1980s, the capacity would increase to about 150 000 tonnes per year. This would meet even the needs of the higher projection up to the year 2000. Therefore, if enrichment is to come under international management, agreement will have to be reached with the operators of the present enrichment plants.

VIII. Fast breeders

The continuation of nuclear power beyond the year 2015 (or beyond the year 2043 if the speculative uranium resources materialize) will necessitate the introduction of the fast breeder. Most fast breeder programmes are based on the uranium–plutonium cycle. A projection has been made that 50 GW(e) of fast breeders will be introduced by the year 2000 and that a rapid increase in this power will ensue, determined by the availability of plutonium from thermal reactors.

From the proliferation risk point of view, the important factor is the high rate of plutonium turnover in the breeder. For the same amount of electricity generated, the amount of plutonium to be reprocessed is about 10 times higher than for thermal reactors. Thus, a 1 GW(e) liquid metal fast breeder reactor (LMFBR) would require an annual reprocessing of 27 tonnes of material containing 2.35 tonnes of plutonium (*Review of Modern Physics*, 1978), while in a PWR the same quantity of material contains only 0.24 tonnes of plutonium.

To put it in perspective, if the high projection for the year 2025 (4 300 GW(e), which would still be only 16 per cent of the total energy consumption) were to be met entirely by fast breeders, the amount of plutonium that would have to be processed *annually* would be about 10 000 tonnes, which is much more than the *total* amount of plutonium that could be accumulated by using the estimated uranium resources. Further expansion of nuclear power would mean a corresponding increase in the amount of plutonium to be processed annually.

In a plutonium economy, fast breeder reactors are likely to remain under national control. This means that colossal quantities of plutonium would have to be transported to international reprocessing plants. The international, or even multinational, management of plutonium under a fast breeder régime would most probably face insurmountable problems in its attempt to minimize proliferation risks.

References

IAEA (International Atomic Energy Agency), 1976. *Directory of Nuclear Reactors*, vol. X, (IAEA, Vienna).

IAEA (International Atomic Energy Agency), 1978. *World Uranium Resources* (IAEA, Vienna.)

IAEA (International Atomic Energy Agency), 1979. *Power Reactors in Member States, 1979 Edition* (IAEA, Vienna).

INFCE (International Nuclear Fuel Cycle Evaluation), 1980a. Working Group 4/INFCE/PC/2/4 (IAEA, Vienna).

INFCE (International Nuclear Fuel Cycle Evaluation), 1980b. Summary volume INFCE/PC/2/9 (IAEA, Vienna).
Nuclear Engineering International, 1980. Nuclear station achievements, annual review–1979. **25**, March.
Reviews of Modern Physics, 1978. 50 (no.1, Part II): S152, January.
SIPRI (Stockholm International Peace Research Institute), 1979. *World Armaments and Disarmament, SIPRI Yearbook 1979* (Taylor & Francis Ltd, London).

Paper 3. Nuclear fuel cycle internationalization: the uncertain political content

G.I. ROCHLIN*

Institute of Governmental Studies, University of California, Berkeley, California 94720, USA

I. Internationalization and the second NPT Review Conference

Following a two-year hiatus for the International Nuclear Fuel Cycle Evaluation (INFCE) to run its course, the commercial use of plutonium will soon return to the international agenda. Five years of intense domestic and international debate have produced no political or technical 'solutions' either to the desire of some states to proceed with plutonium use or to the fears of others about the increased risk of proliferation, diversion or theft that would result from its widespread use. It is unlikely that there will be a return to pre-1974 assumptions of free international trade and autonomous development regulated primarily by multilateral treaties and by International Atomic Energy Agency (IAEA) safeguards, for that would be politically too costly for some key states. It is equally unlikely that others will forgo their intention to proceed with plutonium use. International management or control of sensitive materials or technologies is being proposed as an institutional solution to this dilemma. But the price of negotiability may be a formal division of access rights between the technically advanced countries and the rest of the world.

The institutional details of any such suggestion matter less than the political content. Agreements can be designed to exert broad and strict controls over materials and technology or as a façade to legitimate their further use and trade, to promote plutonium use or to discourage it. What appears most likely at present is that proposals will be structured to permit continued development of plutonium-based fuel cycles by the nuclear weapon states (NWS) only, or by a consortium of the NWS, European Atomic Community (Euratom) and Japan, while hindering or restricting parallel

---

*The opinions expressed in this paper are those of the author and do not necessarily reflect the official policy of his professional affiliation.

development in other countries (Rowen & Wohlstetter, 1979; Lönnroth, 1979). The potential for conflict at the second NPT Review Conference is real and great.

For the USA and others, whose determination to discourage early use of plutonium appears to be wavering in the face of determined opposition, internationalization could satisfy domestic demands for increased control without much restricting the behaviour of the Euratom states and Japan, who are valued political and military allies. West European states and Japan could continue to push for the fuel cycle use of plutonium and early deployment of the fast breeder reactor (FBR) by offering real or symbolic concessions on export and trade policies without otherwise restricting domestic behaviour or trade among themselves.

Internationalization offers a way to institutionalize these divisions without formally acknowledging the political cleavage they entail. For example, agreements for the control of plutonium would probably be applied only to those stocks of material in 'excess' of immediate 'demand'. If institutional rules define demand in terms of FBR development, a *de facto* two-tier system will be in effect for at least one decade and probably much longer. Similar arguments can be made for other aspects of the nuclear fuel cycle (Rochlin, 1979). It remains to be seen whether such agreements will be perceived as merely bowing to present realities or as an attempt to perpetuate them, and whether they would be acceptable to the wider community of states party to the Non-Proliferation Treaty (NPT) in either case.

II. A brief historical review

In the period 1968–70, when the NPT was being drafted and ratified, concern over the use of plutonium fuels was relatively low.[1] Expectations of the future growth of nuclear power were great, and the plutonium-fuelled FBR was seen as a logical follow-on to the uranium-fuelled thermal convertor light water reactor (LWR) and others then just coming into use (Gilinsky, 1977). It was assumed that future expansion would be the task of autonomous domestic industries. The risk of national diversion to military purposes was to be reduced by political agreements and by the application of redefined IAEA safeguards as specified in the NPT.

Although not entirely unforeseen, the Indian nuclear explosion was a shock. Led by the USA, some nuclear exporters began a critical re-examination of their previous acquiescence to transfers of plutonium and

[1] Opinions to the contrary were clearly in the minority, but they were not totally absent. Most notable was Beaton (1977). See also Gilinsky (1977).

reprocessing technology. This concern was fuelled by proposed exports of reprocessing technology by France (to Pakistan, South Korea and Taiwan) and FR Germany (to Brazil, along with enrichment), and fanned by the newly heightened awareness of the additional risk of diversion or theft by subnational groups.

Initial response was muted. In late 1974, 10 nuclear exporters at last submitted to the IAEA a 'trigger list' of materials and facilities not to be exported without prior safeguards arrangements. At the same time, a number of these states met secretly in London to discuss harmonization of export policies.

The first NPT Review Conference in 1975 took place as these events were developing, and the ensuing bitter debate over technical restrictions came close to deadlocking it (Epstein, 1975). Many non-nuclear weapon states (NNWS) party to the NPT took these actions to be a violation of its spirit, which they took as allowing, if not requiring, assistance with indigenous development of all aspects of the commercial fuel cycle.

Although the London Suppliers Group was unable to agree on the application of 'full-scope' NPT safeguards or to a formal embargo, no additional transfers have been proposed since 1976. Of the original proposals, only that of FR Germany to Brazil remains in effect. The USA, as leader of the group of countries seeking further controls has since taken a series of actions to limit further the spread and use of plutonium technologies. The rising tension this caused was quieted, but not dispelled, by the convening of INFCE to study technical and institutional alternatives to the historically conceived LWR-to-FBR development of the fuel cycle.

A second thread of response was the suggestion, again based to some extent on US initiatives, that the IAEA study possible multinational control of fuel reprocessing. After some discussion, the IAEA assembled an expert team to study the economic and technical potential of regional fuel cycle centres (IAEA, 1977). But the IAEA conclusions about economies of scale, the promotion of technical development and the desirability of fuel reprocessing in general were already being challenged before the report was released (Chayes & Lewis, 1977). By that time the USA and others had concluded that the net effect of such a centre might be to accelerate the diffusion of the technology rather than retard it (Menderhausen, 1978).

Nevertheless, the possible internationalization of certain nuclear fuel cycle activities is firmly on the international agenda once more. Although originally focused on reprocessing and mixed-oxide fuel fabrication, these suggestions were extended to encompass management of stocks of separated plutonium and unprocessed spent fuel, as well as multinational participation in uranium enrichment and even nuclear-waste disposal (Petty & Yokota, 1978). And, in an effort to remove some of the incentive for the early use of plutonium, the idea of supply assurance for uranium fuel, including the possible development of an international fuel bank, has been gaining currency.

All of this by no means implies a return to the original Baruch Plan idea of an international nuclear authority. Some ordering of independent behaviour, or its subordination to outside rules, is implied. But these discussions are aimed at the harmonization and co-ordination of independent nuclear policies rather than at their replacement by any superordinating authority. Whether or not these developments violate the 'spirit' of the NPT, they certainly mark a change from the international context in which it was drafted and came into force. Moreover, they will provide a quite different political framework at the second Review Conference for the raising of the unresolved disputes that remain from the first.

III. Motives for early use of plutonium fuels

Fuel-supply arguments are based primarily on allocation, distribution and political control (Neff & Jacoby, 1979). At currently projected growth rates for installed commercial nuclear power on a world-wide basis, net uranium resources appear to be sufficient to fuel even present-design thermal convertor reactors, built by the end of the year 2000, throughout their lifetimes (OECD, NEA & IAEA, 1977). Increased efficiency, continued decrease in growth rates and expected enlargement of the resource base through further exploration would considerably postpone the date of potential scarcity of economically recoverable uranium. The absolute resource stretch obtainable through the recycle of plutonium to thermal reactors can be as high as 30–40 per cent, but no more than perhaps a 10 per cent reduction in uranium demand would be realized during this century (OECD, 1977).

Resource-scarcity arguments for plutonium are most convincing when addressed to the next generation of reactors, whether breeders or advanced thermal convertors (Hebel *et al.*, 1978). But this is a long-term rather than an immediate justification. Since no more than a handful of these will be built before the year 2000, the associated demand for plutonium will remain far smaller than its rate of production. Recycle to thermal reactors may be sought if plutonium is already separated, but is not a convincing argument for its separation.

A second technical argument, the facilitation of radioactive-waste management, is less of a controversy than it was a few years ago. Several reports have de-emphasized waste management as a motive for reprocessing (OECD, 1977; Hebel *et al.*, 1978). Although few, if any, states now seem willing to dispose of the plutonium in spent fuel, there is a growing consensus that extended storage, if properly done, is radiologically and environmentally safe for at least a decade or two.

There are no unambiguous technical, economic or resource criteria for the early use of plutonium. Such pressures as now exist for its separation and use are almost entirely social and political, arising from a combination of domestic and international policies (Rochlin, 1979). Some countries have adopted legal obligations for waste disposal that are interpreted to require reprocessing of spent fuel. Others seek a symbolic resource independence, even though autarky will be impossible until many decades after FBRs are introduced. Some argue that fuel assurance is a problem even if uranium resources are adequate and that any reduction in dependency is a net gain. A few of the most advanced states are competing for early development of an FBR as much to enhance their trade position as nuclear exporters as for domestic reasons.

Prestige is an important factor. For some exporters of high technology, completion of the fuel cycle is seen as advancing their prestige even without the FBR. Less advanced states who aspire to be exporters often make the same arguments or see their prestige and position as linked to reduced dependency on others for technology and equipment. And there are at least a few states that would exploit the ambiguity between peaceful and non-peaceful purposes to seek other forms of prestige and political gain. Technical and economic arguments are the tools of this debate, not its substance.

IV. The improbability of technical alternatives

Two sets of events that might have contributed to delaying or removing the demand for commercial use of plutonium in the next few decades instead appear to have reinforced the historical assumption of the transition from present-design uranium-fuelled thermal reactors to plutonium-fuelled FBRs.

The INFCE exercise was begun largely in response to a series of US studies about technical alternatives and possible increased efficiency for the LWR. Once begun, however, INFCE was reduced to a technical exercise rather than a negotiation. This led to a tendency to aggregate resource and fuel flows, thus disguising the political problems of access and distribution. INFCE working groups were staffed largely by technical experts, for the most part under the watchful eye of governments who saw no advantage to retooling and retraining for alternative reactor designs or alternative fuel cycles. INFCE is thus almost certain to become a gloss on the traditional view that the best strategy is to rely on moderately improved variants of present uranium-fuelled thermal-reactor designs for the time being and to invest further research and development efforts on the plutonium-fuelled FBR, which is the best understood and most extensively studied alternative (Rowen & Wohlstetter, 1979).

The second set of events, the continued downward trend in LWR orders and projections of future growth, appears to have provided an opportunity for reconsideration of reactor design. However, it has also markedly reduced the industry's capacity for innovation. The nuclear industry is in a state of severe recession (if not imminent collapse) in many countries, and neither incentive nor venture capital for new designs exists (Lönnroth & Walker, 1979). Thus, the costs sunk in research and development of the liquid-metal-cooled, plutonium-fuelled FBR are unlikely either to be written off or to be duplicated.

Technical modifications to the historically assumed sequence from uranium-fuelled LWRs to plutonium-fuelled FBRs, possibly via use of plutonium in LWRs, are therefore no more probable than they were in 1975. Fuel cycle options and problems remain technically unchanged. The technical and safeguards situation with which we are faced remains one of potential excess stocks of separated plutonium in the tens of tonnes and potential stocks in the hundreds of tonnes stored in spent fuel. The important questions are where these materials are to be stored, in what form, under whose supervision, with what safeguards and under whose control. These decisions and others will derive from the confluence of domestic and international policies discussed in the next two sections, as shaped by the political understandings that emerge from INFCE.

V. Domestic policies[2]

North and Central America

The USA was the early leader in the promotion of plutonium fuel cycles, having attemped to commercialize the first fast breeder reactor (Fermi 1) and the first commercial reprocessing plant. But plans to complete the full commercial-scale Allied-General Nuclear Services plant in Barnwell, South Carolina, were halted following the October 1976 decision by President Ford to delay plutonium use, pending a review of proliferation risks. President Carter announced a formal deferral of plans to reprocess and recycle plutonium in April 1977, along with a restudy of the domestic FBR programme, an embargo on exports of enrichment and reprocessing technologies and the suggestion for the creation of INFCE. Hearings on the use of reactor-produced plutonium were terminated in 1978, and the USA will be restricted to uranium fuelling for the time being. US waste-management plans are accordingly being altered to include spent fuel as

[2] An overall adaptation from Rochlin (1979). A further useful discussion specifically concerning European co-operative ventures can be found in Menderhausen (1978). A European FBR co-operation can be found in Gray *et al.* (1978).

waste. The FBR programme continues to be a source of internal conflict, effectively removing the USA from the race for early commercialization.

Canada has relied entirely on its own natural-uranium-fuelled, heavy-water-moderated reactor (CANDU). No FBR programme is currently planned. Reprocessing has been given only moderate study due to the low economic value of residual fissile materials in CANDU fuel. However, recent interest in the efficiency of plutonium-based CANDU cycles could eventually result in reconsideration. Present debate in Canada as to whether spent fuel should be stored retrievably, or disposed of, is as yet unresolved.

Mexico has an extensive programme planned, with the mix of LWRs and CANDUs still to be determined. Although the reprocessing decision will not be made until the programme is further developed and clarified, Mexico has announced its intention to construct pilot reprocessing facilities and begin training personnel.

South America

The basis of the contract between Brazil and FR Germany was the acquisition of a complete fuel cycle, including both enrichment and reprocessing. The timetable remains uncertain and may be delayed somewhat by reductions in the scale of the programme; spent fuel disposition plans are unclear. Argentina had a laboratory-scale reprocessing facility and intends to continue experimentation with the technology, but will have little incentive for commercialization if present plans to adopt the CANDU continue. Argentina has also announced its intention to serve as a source of technology for others and at present has such an agreement with Peru.

Western Europe

Although the European Community failed to adopt a formal resolution owing to internal controversies, recommendations from the Commission to the Council have firmly endorsed early reprocessing, both on resource grounds and for future development. Individual programmes, however, vary widely. At one end of the spectrum, FR Germany, France and the UK are partners in the United Reprocessors GmbH (URG) consortium, intended not only to guarantee a reprocessing capacity for the partners but to provide service to outside customers. At the other end are Eire, Luxembourg, Norway and now Austria, all of which have no current nuclear programme at all. The Netherlands, which is a technology supplier, has only a small programme at present.

Following the Windscale Inquiry, British Nuclear Fuels, Ltd (BNFL), one of the URG partners, is prepared to proceed with the construction of a large commercial plant for spent oxide fuels. In addition to domestic

Magnox and Advanced Gas-Cooled Reactor (AGR) fuel, BNFL has a contract to reprocess spent LWR fuel for Japan. The UK has expressed little interest in the recycle of plutonium to present reactors—its primary interest is in support of its FBR programme. Extended storage of spent Magnox fuel is held to be undesirable, but AGR or other spent fuel could be stored for long periods, as high-level wastes must be, until some resolution of the disposal problem is achieved.

Compagnie Générale des Matières Nucléaires (Cogema), the French fuel cycle consortium, is also a URG partner. Its commercial oxide fuel-reprocessing plant at La Hague is just coming into operation. In addition to reprocessing of domestic fuel, Cogema has actively sought foreign service contracts. Customers include FR Germany, Italy, Japan, Sweden and Switzerland. Wastes will be glassified by the Marcoule process. Those of domestic origin will be disposed of within France at an as yet unselected site. Plutonium use is expected to be confined to an ambitious FBR programme.

FR Germany had planned to open its facility as part of a complete fuel cycle and waste disposal complex at Gorleben in the 1980s, but plans are uncertain following the decision of the government of Lower Saxony not to issue the first construction permit. Plans at the moment include away-from-reactor storage of spent fuel not covered by Cogema contracts. Deutsche Gesellschaft für Wiederaufarbeitung von Kernbrennstoffen mbH (DWK), the West German processing combine, has declared its intention to reprocess only fuel of West German origin in any case. FR Germany has had an ambitious FBR programme, but will recycle plutonium in LWRs as required to 'prevent accumulation of stockpiles'.

The only other European government with an immediate interest in the use of reactor-produced plutonium is that of Belgium, which has a long-standing experimental mixed-oxide LWR fuel programme and is co-operating with others on FBR development. Belgonucléaire is one of the most experienced fabricators of plutonium fuels in the world and has contracts with FR Germany and others. There is also some interest in reopening the old Eurochemic plant for reprocessing of domestic fuels. The future of the nuclear programme in Italy, which is also co-operating in FBR work (as is the Netherlands), remains uncertain.

Finland, Spain, Sweden and Switzerland all intend to have spent fuel reprocessed by Cogema and, as a condition for further reactor licensing, are bound to a varying extent to having a 'solution' to the waste problem. None have current plans for plutonium use, and all are being forced to consider extended spent fuel storage as a hedge against short-term shortages of reprocessing capacity. Yugoslavia plans to return spent fuel to the USA.

The USSR and the Council for Mutual Economic Assistance (CMEA)

The USSR is actively pursuing fuel reprocessing both for its FBR programme and for the spent fuel returned to it by fuel-leasing agreements with Finland and the Council for Mutual Economic Assistance (CMEA) countries. The extent and timing of the development of purely commercial oxide fuel facilities are still uncertain, as are waste management and spent fuel storage plans. The USSR is planning early use of plutonium, but the terms of lease to its CMEA partners stipulate only the return of wastes from the spent fuel.

Africa and the Middle East

In these regions, only South Africa is said to be contemplating domestic reprocessing or the use of plutonium fuels at present. Iran's ambitious programme has been severely cut back, although at one time there were plans for a pilot reprocessing facility. Israel and Egypt have discussed a variety of contracts with the USA, all of which would stipulate the return of spent fuel. Other states in these regions are only beginning to consider nuclear programmes and are far from a domestic decision on spent fuel or plutonium.

South Asia

India has an established capability for fuel reprocessing and plans to continue reprocessing domestic spent fuel. Research for an indigenous FBR is a leading motive, but recycle to thermal reactors has not been ruled out. Pakistan had sought to acquire a small reprocessing plant from France, but for the moment the offer has been withdrawn following the Pakistani refusal to modify it for co-processing.

East Asia and the Pacific Basin

Japan is very active in reprocessing and FBR development and has been fabricating plutonium-bearing LWR fuels for several years on an experimental basis. Reprocessing arrangements are required as a condition of reactor licensing, and some spent fuel has already been sent to BNFL under contract. Plutonium produced at the pilot-scale reprocessing facility at Tokai Mura is being stored in solution under terms of an interim agreement with the USA, pending the outcome of INFCE. Japan remains firm in its commitment to extensive use of plutonium fuels in thermal and FBR reactors.

South Korea and Taiwan had both planned for domestic reprocessing, but both have forgone that option for the time being and will have their spent fuel reprocessed abroad. Indonesia and the Philippines are only beginning their programmes, and it is expected that the USA will arrange for the removal of their spent fuel as well. Australia, although a major exporter of uranium and a major voice in the current safeguards debate, has no plans of its own for nuclear power.

VI. Emerging international policies: development, trade and non-proliferation

This *mélange* of domestic policies has neither a technical nor a political centre. The period 1974–76, which marked the end of US dominance of development and trade outside the CMEA area, also marked the beginning of wide divergences on nuclear power policies generally and FBR policies in particular. The restraints imposed since the London Suppliers Group agreement of 1975 may become institutionalized by reciprocal concessions between uranium suppliers and their FBR-promoting customers. But export competition is growing, and there is now the prospect of a 'third tier' of technology suppliers emerging among the more advanced Third World countries (Dunn, 1979). Political costs of continuing restrictions would be lowered if a more general international agreement were forthcoming, but it is not at all clear what concessions would be needed to gain formal accession to a new level of discrimination not inherent in the NPT.

Agreement is most likely with regard to the retransfer of separated plutonium. At the moment, only Cogema is capable of reprocessing commercial oxide fuels, with BNFL next in line. Large stocks of plutonium will therefore be *de facto* limited to NWS for the time being, although smaller, but militarily significant, inventories will be held in countries such as Belgium, FR Germany, India and Japan. The *status quo* will be maintained for perhaps a decade, as BNFL and Cogema are not expected to begin the return of plutonium much before 1990, and neither Belgium, FR Germany nor Japan will have its own plant before then. There is more than a decade in which to negotiate plutonium management and control, whether by institutional, technical or political means. The IAEA has already undertaken a study of its possible role, but the substantive issue of how authority to permit or deny requests for material will be delegated has not been fully addressed (IAEA, 1978).

Present policies vary widely. It is rumoured that future Cogema contracts will include the right of unilateral denial of return of separated plutonium, although previous contracts did not. BNFL will be bound to return separated plutonium to its customers within five years (Windscale Inquiry, 1978). Although both governments insist that the imposition of further conditions will be considered, this would amount to a renegotiation of existing contracts in some cases (Menderhausen, 1978). Countries, such as Sweden, which are more interested in spent fuel removal than plutonium use, appear willing to yield their rights. Others, such as Japan, which have FBR ambitions, will indubitably insist on return. There is a heated debate over West German contracts with Cogema, based ostensibly on the right of free transfer within Euratom and bilateral FBR co-operation agreements. So long as there remains one reprocessor with many potential customers, Cogema's rights are likely to dominate, but this does not constitute any resolution of underlying disagreements.

Opinion about the need to manage spent fuel stocks is more divided, although spent fuel was also included in the IAEA study. Spent fuel is the 'ore' for plutonium, but separation would not be so simple for subnational groups, and physical security is enhanced by the high levels of radiation. Safeguards against national diversion may also be more readily devised for the large and quite identifiable spent fuel assemblies. Although rather strict conditions on the transfer or reprocessing of fuel originating from Australia, Canada and the USA have been imposed, there is little agreement as to what form of additional control or regulation would contribute significantly to non-proliferation. These differences are exacerbated by widely divergent national policies with regard to waste management and plutonium use.

The most serious differences, however, centre on the further spread of reprocessing, particularly in those cases where it is already linked to a possible FBR programme. Commercial reprocessing is limited to the NWS for the time being, and there is already discussion of unease in FR Germany and Japan over a possible attempt by the NWS to institutionalize the present situation (Lönnroth, 1979). Attempts to allay suspicion of commercial motives by including FR Germany and Japan would stir opposition in Belgium and other Euratom partners in European FBR development. Whether or not all of Euratom were included would make little difference to advanced states such as Sweden that have no FBR programme by choice, but Third World countries are likely to bridle either at the further widening of the NPT-formalized division between the NWS and the NNWS or at the creation of a new division based on present levels of technical development. These problems and others were disregarded while INFCE ran its course, but are sure to be raised again by the time of the second NPT Review Conference. The argument, as to whether reprocessing should be considered an effective transfer of quasi-military capability to be discouraged under Articles I and II of the NPT or a necessary fuel cycle development step to be encouraged and assisted under Article IV, will continue.

VII. Conclusions

The NPT did not impose harmony on international nuclear policies in 1970. Rather, it was the perceived harmony of policies in the mid-to-late 1960s that led to its successful drafting and acceptance. By the time of the first Review Conference in 1975, emerging policy differences were already so serious that deadlock was avoided only by some adroit political manoeuvring by the Chair (Epstein, 1975). Barring an unforeseen and improbable restoration of some modicum of consensus over the next few months, policy differences will be sharper and better articulated than they were five years ago. Prospects for a harmonious outcome are therefore much reduced.

The key questions are political ones: (*a*) who will be allowed access to material and technology and with what restrictions; (*b*) who will govern releases or authorized transfers and under what conditions; (*c*) whether the underlying non-proliferation strategy is promotion and co-optation or control and denial; (*d*) whether national programmes are to be standardized, co-ordinated or forgone, and (*e*) whether membership is to be regional or global and inclusive, exclusive or two-tier?

Internationalization or multilateralization of plutonium or spent fuel stocks, or of reprocessing and waste management, will not automatically resolve these differences. Rather, they must be deliberately encompassed (see Papers 14 and 21). Past studies on possible modes of internationalization have generally avoided such trying matters and turned instead to discussion of legal instruments and institutional structures, thus begging the question as to what political goals these institutions are designed to serve (IAEA, 1977; IAEA, 1978; Fox & Willrich, 1978; Atlantic Council, 1978). For the time being, such proposals are receiving cautious support, but only to the extent that the political ambiguity allows states to assume that their own goals and values will be incorporated (Menderhausen, 1978).

Most of the discussion to date has taken place among the NWS, Euratom members and Japan and appears increasingly to be directed towards removing further sources of conflict among them while responding to domestic concerns over the spread of nuclear weapons to other less advanced states. If the trend continues, proposals are likely to be quite context-specific and difficult to promote as equitable to the community- at-large. They are as likely to be a source of new disagreements at the second NPT Review Conference as a palliative for old ones.

References

Atlantic Council, 1978. *Nuclear Power and Nuclear Proliferation*, Report of the Atlantic Council's Nuclear Fuels Policy Working Group (Atlantic Council, Washington, D.C.)

Beaton, L., 1977. Nuclear fuel for all, *Foreign Affairs*, 45:662–69.
Chayes, A. & Lewis, W. (eds), 1977. *International Arrangements for Nuclear Fuel Reprocessing* (Ballinger, Cambridge, Massachusetts).
Dunn, L.A., 1979. Half past India's bang, *Foreign Policy*, 36:71–89.
Epstein, W., 1975. *Retrospective on the NPT Review Conference: Proposals for the Future* (The Stanley Foundation, Muscatine, Iowa).
Fox, R.W. & Willrich, M., 1978. *International Custody of Plutonium Stocks : A First Step Towards an International Régime for Sensitive Nuclear Energy Activities* (International Consultative Group on Nuclear Energy; The Rockefeller Foundation, New York, and the Royal Institute of International Affairs, London).
Gilinsky,V., 1977. Plutonium, proliferation and policy, *MIT Technology Review*, February: 58-65.
Gray, J.E., *et al.*, 1978. *International Cooperation on Breeder Reactors* (The Rockefeller Foundation, New York).
Hebel, L.C. *et al.*, 1978. Report to the American Physical Society by the Study Group on Nuclear Fuel Cycles and Waste Management, *Reviews of Modern Physics,* 50(1): part 2.
IAEA (International Atomic Energy Agency), 1977. *Regional Nuclear Fuel Cycle Centers,* Report of the IAEA Study Project (IAEA, Vienna).
IAEA (International Atomic Energy Agency), 1978. *International Management of Plutonium and Spent Fuel,* prepared by the Secretariat with the assistance of expert consultants (IAEA, Vienna).
Lönnroth, M., 1979. *The Politics of the Back End of the Nuclear Fuel Cycle in Sweden* (Secretariat for Future Studies, Stockholm).
Lönnroth, M. & Walker, W., 1979. *The Viability of the Civil Nuclear Industry* (International Consultative Group on Nuclear Energy; The Rockefeller Foundation, New York, and the Royal Institute of International Affairs, London).
Menderhausen, H., 1978 *International Cooperation in Nuclear Fuel Services: European and American Approaches,* Rand Corporation Report P-6308 (The Rand Corporation, Santa Monica, California).
Neff, T.C. & Jacoby, H.E., 1979. Non-proliferation strategy in a changing nuclear fuel market, *Foreign Affairs,* 57:1123–43.
OECD (Organisation for Economic Co-operation and Development), 1977. *Reprocessing of Spent Nuclear Fuels in OECD Countries, A Report by an Expert Group of the OECD Nuclear Energy Agency* (OECD, Paris).
OECD (Organisation for Economic Co-operation and Development), NEA (Nuclear Energy Agency) & IAEA (International Atomic Energy Agency), 1977. *Uranium: Resources, Production, and Demand, 1977* (OECD, Paris).
Petty, G.M. & Yokota, M., 1978. *International Nuclear Service Centers: A Bibliography,* Rand Corporation Report P-5930 (The Rand Corporation, Santa Monica, California).
Rochlin, G.I., 1979. *Plutonium, Power, and Politics: International Arrangements for the Disposition of Spent Nuclear Fuel* (University of California Press, Berkeley and Los Angeles, California).
Rowen, H. & Wohlstetter, A., 1979. *U.S. Non-Proliferation Strategy Reformulated,* Draft report to the Council on Environmental Quality, the Department of Energy, and the National Security Council (CEQ, Washington, D.C.).
Windscale Inquiry, 1978. *The Windscale Inquiry: Report by the Hon. Mr. Justice Parker.* (HMSO, London).

Paper 4. An international plutonium policy

A.R.W. WILSON*

Australian Atomic Energy Commission, P.O. Box 41, Coogee, N.S.W. 2034, Australia

I. The plutonium dilemma

Preparations for the introduction of thermal plutonium recycle and fast breeder reactors into national nuclear power programmes face society with a grave conflict of interests. On the one hand, they promise a significant contribution to energy supplies at a time when the future adequacy of supplies is uncertain; on the other hand, they threaten to heighten the risk of nuclear weapon proliferation.

This paper discusses technical and institutional measures which might reduce the proliferation risks of commercial plutonium separation and recycle and suggests an international plutonium policy for constraining the proliferation risks of thermal plutonium recycle and fast breeder reactors.

II. The plutonium proliferation risk

So long as the production of a nuclear weapon by a non-nuclear weapon state necessarily involved the separation of plutonium from diverted fuel, it was possible to assume that reasonably prompt detection of the diversion by safeguards would allow time for the international community to try to dissuade the state concerned from proceeding before it achieved its nuclear weapon objective. If non-nuclear weapon states gain access to separated plutonium, such an assumption will no longer be possible, since the fabrication of separated plutonium into a nuclear weapon is a fairly easy task (ERDA, 1976).

---

* The opinions expressed in this paper are those of the author and do not necessarily reflect the official policy of his professional affiliation.

III. Alternative fuel cycles

If it were possible to establish that an alternative, more proliferation-resistant fuel cycle offered equivalent or superior economic and resource utilization benefits, there might be a chance of diverting interest from plutonium recycle. Exploration of this possibility is one of the tasks assigned to the International Nuclear Fuel Cycle Evaluation (INFCE). On the assumption that informed and deliberate technical and economic decisions have determined the direction in which the nuclear industry is currently proceeding, one cannot very highly rate INFCE's chances of finding attractive alternative fuel cycles. In any event, it seems unlikely that states with major research and development and capital investments in the uranium–plutonium fuel cycle could be persuaded to abandon that fuel cycle in favour of a much less developed and tested concept.

IV. Process-design modifications

As an insurance against the possibility that it will be unable to establish that there is a viable proliferation-resistant alternative fuel cycle, INFCE is also looking at a number of process-design modifications which might be helpful in upgrading the proliferation resistance of fuel cycles involving separated plutonium. They include measures for reducing the presence of separated plutonium in the fuel cycle, such as co-location, co-conversion and co-processing, and measures for increasing the difficulty of handling recycled nuclear materials, such as pre-irradiation, spiking with radioactive material and partial processing. Unfortunately, in general they appear to be more relevant to the risk of subnational diversion than to the possibility of diversion by states. The non-proliferation value in making their adoption a condition of supply in bilateral agreements would have to take account of the economic burden which would be incurred by the operator and the degree to which this might alienate his support of non-proliferation arrangements.

V. Operating restrictions

The proliferation risk of national programmes for the commercial separation and recycle of plutonium could also be constrained by operating restrictions on the stockpiling of plutonium, the throughput of reprocessing plants, fuel irradiation times and other proliferation-related operating parameters. By limiting the amount of plutonium in stockpiles and ensuring that the stockpiled material was in a form as little suited to weapon use as possible, such restrictions might make the diversion of stockpiled plutonium a less attractive proposition for a potential proliferator. It is unlikely that such restrictions would be acceptable as supply conditions in bilateral agreements, and their widespread adoption could possibly only be achieved as a component in a treaty which provided compensatory benefits.

VI. Multinational and international control

If it is found that high-proliferation-risk nuclear activities, such as the commercial separation and use of plutonium, are essential to meet energy needs, then from a non-proliferation viewpoint, it is desirable that they proceed under the control of the nuclear weapon states or groups of states having no credible motivation for attemping to divert the activities to nuclear weapon production. Monopolistic control of plutonium fuel cycle activities by the nuclear weapon states would undoubtedly be unacceptable to most non-nuclear weapon states. However, a number of non-proliferation suggestions put forward to date have envisaged some form of multinational or international control of fuel cycle activities. Such control could be achieved at either the host-government level or the plant level.

Control at the host-government level would involve the creation of multinational or international enclaves within which industry or governments could construct, own and operate fuel cycle facilities on a multinational or a purely national basis. Multinational or international control of the enclaves would minimize both the covert diversion and the plant seizure proliferation risks, while allowing industry and governments to develop fuel cycle activities according to their own capital investment and operating-efficiency perceptions. The political difficulties which would be involved in creating multinational or international enclaves make it unlikely that this concept could be realized in time to cope with expected industrial developments.

Control at the plant level would involve multinational or international governmental investment in, and management of, plutonium fuel cycle facilities located on national territory and operated subject to an agreement concluded between the host government and the multinational or international grouping. Such an arrangement would minimize the risk of covert diversion and reduce, although not eliminate, the risk of plant seizure. Its non-proliferation credibility would be dependent, *inter alia,* on the spectrum of states participating in the project, the perceived political stability and non-proliferation credentials of the host state, and the governmental grouping's relationship with the host state. The multinational or international project group could be structured on a variety of existing institutional models, according to the intended degree of multinational or international involvement in policy determination, management decisions and operations.

Although multinationalization of fuel cycle facilities is generally perceived as requiring multinational governmental participation in their control, it has been suggested that worthwhile non-proliferation benefits would accrue also from multinationalization achieved through multinational commercial and industrial equity investment in fuel cycle facilities (Atlantic Council of the United States, 1978). Multinational equity investment which provided for multinational participation in the management of the facilities would ensure more openness about their operation and to that extent enhance the effectiveness of safeguards. However, it would not be nearly so effective as multinational governmental participation in protecting the project against the possibility of takeover of its operations by the host government. Nevertheless, because it is a much more readily achievable arrangement, there could be merit in adopting it as an interim international policy objective for all future fuel cycle plants.

Multinationalization and internationalization of the various plutonium fuel cycle activities posing major proliferation risks could proceed progressively through individual initiatives tailored to meet the particular technical, industrial and political circumstances of the activity. The reprocessing and storage of separated plutonium are obvious and promising candidates for early attention. At the other end of the spectrum there are some activities involving plutonium, such as reactor operation and internal fuel element transport, which would be impractical to remove from national control. Thus, multinationalization and internationalization should not be seen as adequate in themselves to contain the added proliferation risks which would flow from the adoption of thermal plutonium recycle and fast breeders.

International plutonium storage

International storage of plutonium is aimed at making the sudden diversion of stockpiled plutonium a more physically difficult and politically dangerous operation. Ideally, the plutonium should be stored outside of the territory of the owner state. Where this is not practical, the physical barrier to diversion would probably be relatively trivial, although international responsibility for storage would still raise the political costs involved in a state forcibly resuming control of the plutonium it had deposited with the store. It might also promote international confidence regarding a states's intentions by making its plutonium-related activities more visible.

The non-proliferation value of any scheme for the international storage of plutonium will be very much dependent upon the agreed conditions of access. While some plutonium stocks may not be required until the commercial introduction of plutonium recycle and fast breeders, other stocks may be regarded by their owners as stocks which should be available to them on short notice. If it turns out that to meet the latter situation all stocks in the international store must be regarded as available on short notice for any peaceful uses nominated by the owner, international storage may not provide any greater non-proliferation reassurance than existing safeguards arrangements. It would then be necessary to consider whether such an international storage scheme might not be open to exploitation by states that wished to legitimize efforts to develop national nuclear programmes embracing high-proliferation-risk activities.

A number of states meeting under the aegis of the IAEA are already examining the problems involved in the international storage of plutonium.

Multinational reprocessing plants

The scope for multinationalizing or internationalizing reprocessing operations has attracted considerable attention not only because of the potential for reducing proliferation risk, but also because multinationalization or internationalization might offer participating states both economies of scale and improved supply assurance. The desirability of rationalizing spent fuel transport operations and a perceived identity of nuclear interests among states in particular geographic areas have led to suggestions that priority should be given to the creation of regional multinational reprocessing plants.

The establishment of regional multinational reprocessing facilities would involve capital investment by governments and agreement on complex arrangements for their construction and operation. It has been suggested that a more practical means of achieving multinational control of reprocessing services at the regional level would be through the creation of purely administrative regional nuclear supply agencies (Wilson, 1975). If the participating member states were to agree that the regional nuclear supply agency should serve as the exclusive channel through which they would seek access to established reprocessing services, the agency could use its monopoly position both to obtain the best commercial terms for its members and to ensure the application of adequate safeguards.

VII. The recycle-breeder time-scale

Most countries base their case for contemplating plutonium recycle on resource considerations. It is not, as yet, established that there would be any economic benefit in moving to thermal plutonium recycle or fast breeders should uranium supplies remain adequate and uranium prices remain stable.

Thermal recycle of plutonium in light water reactors promises substantial uranium savings, while plutonium recycle in fast breeders offers the possibility of reaching a situation of virtual independence of uranium supplies. A reduced dependence on uranium supplies could be of considerable strategic and economic importance if a uranium supply shortage develops or the price of uranium rises appreciably. Thus, the rate of adoption of plutonium recycle in both thermal reactors and fast breeders will be strongly influenced by national perceptions of the adequacy of the uranium resource base and the ability of the uranium mining industry and market to produce and distribute the uranium supplies needed to meet demand.

Assessments of the time period over which achievable uranium production, based on known resources, will prove adequate to meet demand vary widely, mainly because of differences of view concerning the likely rate of growth of installed nuclear generating capacity. However, there seems to be agreement that if a uranium shortage does develop it will not do so until after the year 2000 (Martin, 1979). Thus commitment to power programmes based on thermal plutonium recycle and fast breeder reactors would seem to be avoidable without significant penalty until towards the end of the century if storage of spent fuel from the once-through operation of light water reactors is accepted as an interim, if not a longer-term, measure.

VIII. The plutonium challenge

Even though commitment to thermal plutonium recycle and fast breeders can be deferred, it would probably be generally accepted that continued technical development of thermal plutonium recycle, and particularly of fast breeder reactors, is necessary as an insurance against the possibility that uranium production may be inadequate to meet demand at the beginning of the next century. Thus, not only is it necessary to develop arrangements within which nuclear power programmes utilizing thermal plutonium recycle and fast breeders can proceed without raising unacceptable proliferation risks, but it is also necessary to find a way of allowing the plutonium recycle and fast breeder technical development effort to proceed without prejudicing the willingness of states to participate in new arrangements as they are developed. Even if several years are available for the first task, an answer to the second problem must be found urgently. A plutonium non-proliferation policy which would provide a realistic basis for international action to tackle both the long- and the short-term problems is outlined below.

An international plutonium policy

The first and the priority objective of any plutonium non-proliferation policy must be the maintenance and strengthening of the political fabric of the existing non-proliferation régime. A shared political commitment to non-proliferation is an essential foundation for the effective functioning of all technical and institutional non-proliferation measures. Concern over the use of plutonium must not be allowed to downgrade the importance of such initiatives as the development of a comprehensive test ban treaty, measures to give non-nuclear weapon states greater confidence in their security status, achievement of the widest possible participation in the Non-Proliferation Treaty and the Treaty of Tlatelolco, the encouragement of interdependence in nuclear trade, the rationalization of bilateral non-proliferation arrangements, and the widest possible adoption of full-scope safeguards.

The second objective would be to confine plutonium recycle and fast breeder research and development (R&D) to a few large states which had the technical resources to carry on meaningful programmes and which were willing to commit themselves to bringing their own plutonium activities within international arrangements as such arrangements were developed. The plutonium R&D sanctioned states might be designated internationally on the basis of the extent of their R&D involvement and the adequacy of their non-proliferation credentials. It would then be up to supplier states to key the export conditions in their bilateral agreements, and particularly the exercise of prior-approval reprocessing and retransfer rights, to the concept of plutonium R&D-sanctioned states. States excluded from plutonium-related R&D activities would need to know that the technology developed by sanctioned states would be made available to them within internationally agreed conditions and on reasonable terms if and when its use became economic. To this end the group of plutonium R&D states should be seen to contain credible technology-exporting competitors—a requirement which would probably call for active US involvement in plutonium R&D. It might also be helpful if suppliers were to announce that, in exercising their reprocessing and retransfer prior-approval rights, they would consider whether or not there was evidence that the plutonium R&D-sanctioned states were meeting their obligations regarding availability of developed technology and acceptance of international arrangements for plutonium activities.

The third objective of the policy would be to increase the confidence of states that in accepting a moratorium on their preparations for adoption of plutonium thermal recycle and fast breeders they would not be jeopardizing the continuity of their nuclear power generation programmes. This would require a concerted effort by both suppliers and consumers to develop and implement measures which would provide greater assurance of long-term supply of uranium and enrichment services. Useful measures might include the rationalization of governmental supply agreements, the creation of emergency back-up machinery, such as fuel banks, improvements in market intelligence, and the involvement of customers in supply operations. If the problems posed by waste disposal could be overcome, nuclear weapon states might even be willing to offer long-term fuel-leasing arrangements on attractive terms to non-nuclear weapon states which undertook not to embark on plutonium-related activities. Measures to ensure the continued availability of adequate supplies of uranium and enrichment services at stable prices would also serve to defer the time when the adoption of plutonium recycle and fast breeders becomes economically attractive.

A fourth objective of the strategy would be to achieve international agreement on the technical and institutional arrangements within which programmes for the commercial separation and use of plutonium might go ahead without raising unacceptable proliferation risks. Agreement on the international storage of plutonium would have to be followed by the development of arrangements for minimizing the proliferation risk of other fuel cycle activities involving separated plutonium, such as conversion and fuel fabrication, and the specification of the technical and operating controls and safeguards to be applied to national programmes.

The fifth objective would be the early development and realization of an effective and acceptable form of multinational control of reprocessing operations. The early availability of reprocessing capacity, controlled and operated according to internationally agreed arrangements, would undermine the credibility of claims that the construction of national reprocessing plants was essential because of the unacceptability of interim storage of spent fuel. From a non-proliferation viewpoint, it would be desirable that the capacity be provided by way of the transfer of existing plants from national to multinational control. Transfer of existing plants might also help to avoid excess capacity problems. Access to such multinationally controlled reprocessing services should be subject to an agreement to place all recovered plutonium in international storage until required for demonstrable power-production purposes.

The sixth and final policy objective would be the substitution of supplier-imposed restrictions by obligations freely accepted by states in the context of arrangements and treaties developed by all potential parties. As consumer states develop their own technology and as new nuclear material supply sources become available, supplier-imposed non-proliferation restrictions, will inevitably represent less of a barrier to proliferation. Their gradual substitution by obligations undertaken in the context of treaties and multinational arrangements would acknowledge their transient usefulness and remove a source of tension in supplier–consumer relationships. In the longer term it may be advantageous to build all non-proliferation arrangements into a single treaty, so that access to the potential supply benefits, through membership in the treaty, provides an added incentive for states to develop their nuclear power programmes within the various internationally agreed non-proliferation restraints.

References

Atlantic Council of the United States, 1978. *Nuclear Power Nuclear Weapons Proliferation* (Westview Press, Boulder, Colorado), vol.1, chapter 6.

ERDA (US Energy Research and Development Agency), 1976. Proposed statement at AIF/ANS Washington meeting, *Nucleonics Week: AIF/ANS Washington Special,* 16, November.

Martin, D., 1979. INFCE Working Group papers reflect divergent international views, *Nuclear Fuel,* 4(17):9.

Wilson, A.R.W., 1975. Regional nuclear supply agencies—a new non-proliferation concept, October (unpublished).

Paper 5. Internationalization of the nuclear fuel cycle

B. SANDERS*

United Nations, Centre for Disarmament, Room 2755, New York, New York 10017, USA

I. Introduction: the nature of nuclear proliferation

This paper attempts to clarify the role that internationalization of parts of the nuclear fuel cycle might play in reducing nuclear weapon proliferation risks.

From the text of the Non-Proliferation Treaty (NPT) one may deduce that the term 'nuclear weapon proliferation' implies the acquisition, by manufacture or transfer, of nuclear weapons or other nuclear explosive devices, or of the control of such weapons or devices, by any state which had not manufactured or exploded a nuclear weapon or other nuclear explosive device prior to 1 January 1967 (Articles II and IX.3). The NPT also gives some indication of why the proliferation of nuclear weapons is undesirable in stating, as the belief of the parties, that "the proliferation of nuclear weapons would seriously enhance the danger of nuclear war" (NPT preamble). However, the avoidance of nuclear war is obviously not the only reason for following a policy of non-proliferation. Nuclear proliferation has been a continuing phenomenon ever since the United States acquired a nuclear military capacity. When the NPT was concluded, it recognized the existence of five nuclear weapon states. Since then, one more state has tested a nuclear device, at least two are believed to have acquired a nuclear military capacity, and others are said to be close to obtaining it. Yet, since 1945 no state has used nuclear weapons in war.

As the principal international instrument designed to stop nuclear proliferation once and for all, the NPT was logically constrained from recognizing that proliferation might be a continuing process. In categorizing those states which by 1 January 1967 had manufactured and

* The opinions expressed in this paper are those of the author and do not necessarily reflect the official policy of his professional affiliation.

exploded a nuclear weapon or other nuclear explosive device as nuclear weapon states, it implied that no other state could ever become a 'real' nuclear weapon state. Any nation that would detonate a nuclear device after that date would be acting against the precept "thou shall not (nuclearly) proliferate (further)", which was demonstrated by reactions to the detonation by India of a nuclear explosive device in 1974 and the reports that Israel and South Africa possess, or are about to have, nuclear weapons. None of these countries is a party to the NPT. Nevertheless, there appears to be a tendency, also among other non-parties, to condemn such states, and in particular the latter two, for having acted against the non-proliferation precept.

The reason for this reaction seems to be central to a discussion of the nature of nuclear proliferation and the role it plays in international relations. It seems to arise mainly from the fact that, although there are probably few international observers who suspect these states of planning to start a nuclear war in the near future, they have the ability to influence, in the sense of maintaining or changing, the relations among themselves and other states by the very fact that they are assumed to possess one or more nuclear explosive devices or to have the capacity to manufacture such devices and the means to use them.[1] If one accepts this reasoning, the real or supposed acquisition of a nuclear explosive device plays a greater role than that of 'merely' enhancing the risk of nuclear war: it works to disturb international relations in a manner considered to be undesirable by the majority of nations.

Any state that possesses the sophisticated nuclear installations in which highly enriched uranium or plutonium is handled in bulk form may now be thought to have the technological means of making an explosive device. The judgement of whether a state actually is in such a position can be based in part on knowledge of that state's nuclear development and in part on a presumption of its will to use its technological means to exert its nuclear capacity for military purposes. Where such a presumption is not based on proven facts, it may rest on an assessment of a state's political incentives. The extent to which this presumption influences the behaviour of other states is in itself a reflection of the ability of the country in question to exercise a nuclear policy in international relations. In other words, the fear of a country's nuclear capability, based on the presence of certain installations, combined with the presumption of a state's intentions, is an element of proliferation. The intentions may vary and be variously interpreted. In the final

[1] Whether and how a state can "influence the relationship between itself and other states" by nuclear means will depend in part on its own political, military and strategic position and that of the other states concerned. Some states may be able to do so through the possession of a few explosive nuclear devices, deliverable by relatively unsophisticated means. Other countries, particularly highly developed states with powerful conventional forces and/or situated near nuclear weapon states, might be neither able nor presumably willing to influence the strategic situation by their own nuclear weaponry unless they disposed of a formidable arsenal of such devices. This variable will enter into the 'presumption of intention' discussed below.

instance, it is the possession of the nuclear installations which may upset the balance.[2]

It is obvious from 30 years of nuclear application that it is impossible to avoid a situation where no single state beyond the five recognized nuclear weapon states is at least capable of obtaining the means to manufacture a nuclear weapon. A considerable number of states, parties as well as non-parties to the NPT, do have that capacity. Purely technologically, it is a relatively simple step from the possession of fissionable nuclear material, suitable in quality and quantity, to the manufacture of a deliverable nuclear explosive device. If, as indicated above, the possession *per se* of the wherewithal to make such a device may be an element of nuclear proliferation, then many states are in a position to disrupt the non-proliferation régime. To maintain, as some authoritative writers do, that further attempts to contain proliferation are therefore fruitless is a self-defeating line of reasoning because, in practice, the large majority of 'near-nuclear' nations, although technically capable of a military effort, would for various valid reasons prefer to avoid embarking on that path. Encouraging such states to continue in their present policy should be a principal purpose of the non-proliferation régime. In particular, if the assumed intention to make military use of nuclear capacity is tantamount to the actual possession of a military nuclear capability, it is essential to disprove such assumptions in order to prevent an ensuing arms race.[3]

II. Early proposals for control

The first non-proliferation scheme was contained in the Baruch Plan of 1946, which envisaged the creation of an international atomic development authority. It is highly significant that among the functions of this body the first two were the managerial control or ownership of all atomic energy activities potentially dangerous to world security and the power to control, inspect and license all other atomic activities.

---

[2] The fact that nuclear energy serves peaceful as well as destructive purposes needs no discussion here. The nuclear fuel cycle contains several stages at which technology offers a choice between civil and military uses (see, for example, SIPRI, 1975). A recent SIPRI publication (SIPRI, 1979) gives a good indication of 'the dual nature of nuclear fission'. It should be noted that these remarks pertain only to overtly 'peaceful' nuclear facilities. A state intending to acquire a military nuclear capacity by use of a 'dedicated fuel cycle' (that is, a series of installations designed, constructed and operated, possibly clandestinely, for the purpose of manufacturing nuclear weapons) is not considered, since it is unlikely that such a state would join, or even comply with, the obligations of a multilateral non-proliferation régime.

[3] The terms 'state' and 'nation' are used here interchangeably to denote a political and territorial entity recognized as such in international law. Attempts on the part of subversive groups to acquire a nuclear capacity are not discussed since it is assumed that the state concerned would take measures to thwart such attempts.

Thus already at the onset of the atomic age it was proposed to place 'potentially dangerous' activities under international control, whereas other activities, presumably less risky for world security, might be safeguarded by other means. The Baruch Plan and a number of other proposals were discussed for several years. Among them was a Soviet proposal for "measures of control of atomic energy facilities", which would be carried out by an international commission within the framework of the UN Security Council. Like the initial US proposal, the Soviet plan distinguished between control and inspection. Although neither plan was put into effect, it is useful to note that well over 30 years ago the two major powers realized that, to avoid misuse, certain nuclear facilities should be put under international control and that this would not be the same as mere verification.

Although the question of establishing a single international body to control nuclear energy was not seriously raised again, traces of the initial proposals can be found in Article XII A.5 of the Statute of the International Atomic Energy Agency (IAEA), which gives the Agency the right, under particular conditions, to require that special fissionable materials produced are deposited with it in order to prevent stockpiling of these materials.

III. IAEA safeguards

Essentially, the IAEA Statute sets out the so-called safeguards of the Agency essentially as measures of verification, that is, an 'after-the-fact' determination of compliance with an undertaking not to use certain items in a proscribed manner (see, e.g., IAEA Statute Article XII A.6 and C). The IAEA safeguards form a basic element of the NPT, which in Article III.1 requires each non-nuclear weapon state party to the treaty "to accept safeguards, as set forth in an agreement to be negotiated and concluded with the International Atomic Energy Agency in accordance with the Statute of the International Atomic Energy Agency and the Agency's safeguards system, for the exclusive purpose of verification of the fulfillment of its obligations assumed under this Treaty with a view to preventing diversion of nuclear energy from peaceful uses to nuclear weapons or other nuclear explosive devices". The IAEA safeguards system may be applied by virtue of a state's adherence to the NPT or to another multilateral instrument, such as the 1967 Treaty for the Prohibition of Nuclear Weapons in Latin America (see Article 13), or pursuant to another agreement under which the state gives an undertaking not to use any or some of its nuclear material or facilities for a purpose prohibited in that agreement, coupled with the acceptance of measures by which compliance with undertakings may be ensured. It consists of a set of measures (including elaborate bookkeeping and reporting procedures and on-the-spot inspection) whose technical objective is the timely detection of diversion of significant quantities of nuclear material from peaceful nuclear activities to the manufacture of nuclear weapons or of other nuclear explosive devices, or for purposes unknown, and the deterrence of such diversion by the risk of early detection (IAEA INFCIRC/153, paragraph 28).[4]

However, although certainly comprising a strong element of deterrence, safeguards basically remain measures of detection, whether or not an act of non-compliance has taken place.

Strengthening the safeguards régime

Efforts made in the past seven or eight years to strengthen the safeguards system, particularly consultations among supplier nations to harmonize their policies in this respect, have been well publicized (see, for instance, UN, 1977). Talks on improving the international safeguards régime have been held especially in the framework of the so-called London Club of nuclear-exporting states, a number of which now require, as a condition for the export of nuclear material and equipment, that all nuclear activities in the recipient state be covered by IAEA safeguards.[5]

IV. Additional measures

With the increase in the size and complexity of nuclear facilities and in the quantities of fissionable material they process, apprehension has been growing that available safeguards techniques may not always yield the required degree of confidence that the purpose of safeguards has been met.[6]

---

[4] The specific formulation in INFCIRC/153 was initially drawn up with particular reference to safeguards applied pursuant to the NPT, but it is fair to say that the principle underlies the application of IAEA safeguards in general. Some of the terminology used required further definition. In general terms, 'timely' and 'early' indicate that safeguards should be so applied as to ensure that detection of any diversion should occur before the material can be used to manufacture an explosive device. The concept of 'timeliness' is connected with that of 'significance'. The amounts of material of which the diversion is to be discovered in time will vary, depending on the steps required to convert the material into an explosive. The more directly a material may be used for that purpose, the greater the significance will be of even a small amount, and the earlier detection of diversion must be assured.

[5] So-called 'full-scope safeguards' are demanded, *inter alia*, by Australia, Canada, Sweden and the USA. See, for example, the Nuclear Non-Proliferation Act, Public Law 95-242 of 10 March 1978, which governs the US approach in this regard.

[6] INFCIRC/153, paragraph 30, states that "the technical conclusion of the Agency's verification activities shall be a statement...of the amount of material unaccounted for over a specific period, giving the limits of accuracy of the amounts stated". There are nuclear installations where the operator's measurements contain uncertainties which exceed "significant quantities" of the material involved. In such cases, the verifying authority, which may not be the operator, could face a situation where its 'statement' is unsatisfactory in terms of the purposes of the agreement.

This has contributed to the feeling among many observers that it is not enough to rely on safeguards to detect non-compliance with the safeguards régime and that they should be supplemented by various measures of prevention. Such considerations lead increasingly towards a two-directional trend: a search for less proliferation-prone fuel cycles (avoiding, in particular, the use of plutonium as a reactor fuel) and a tendency to impose, wherever possible, various forms of control on the uses of nuclear energy, which go beyond mere verification of those uses.

The INFCE study

Efforts to find alternative nuclear fuel cycles, less prone to add to a nuclear explosive capability, are taking place mainly in the International Nuclear Fuel Cycle Evaluation (INFCE). Even if technological solutions can be found, there may be economic, technological and political obstacles to their introduction. INFCE is likely to yield a number of worthwhile ideas, pointing to a greater need for international co-operation, combined with measures of prevention and control.

Export restrictions

Co-operation among states in the peaceful uses of nuclear energy has recently contained elements of control. Following agreements reached in the London consultations, a number of suppliers now demand that certain supplied items, or special nuclear material produced through the use of such items, not be re-exported without their prior consent. There are also new requirements for approval by the supplier before exported fuel may be reprocessed. In addition, the majority, if not all, of the supplier states now follow a restrictive policy on the export of technology and equipment for specifically sensitive stages of the nuclear fuel cycle. Such restrictions are designed to help prevent certain nuclear uses, but they lack the factor of physical control implicit in the various plans conceived after World War II, in that they cannot prevent a country from taking the unilateral decision to use a nuclear capacity, once achieved, in a proscribed fashion (Szasz, 1979).

Internationalization and control

In the last instance, the assurance that such a decision will not be taken or will not be carried to its ultimate fruition can probably be achieved only if the operation of the facilities where special fissionable material is produced, processed or stored is controlled by an outside authority. Aside from a situation where the nuclear activities of one state take place under the exclusive control of another, such an authority must by definition be international. By internationalization of the fuel cycle is meant submitting the fuel-cycle facilities that handle or contain fissionable nuclear material suitable in quality and quantity for the manufacture of a nuclear explosive device to an international authority which has the physical power to control the use to which such material is put. Internationalization in this sense implies the removal from exclusive national control of nuclear material and facilities, and the assignment to a multinational authority of the power to decide what use should be made of those items, as well as the physical means to carry out that decision. Internationalization might either take place in co-operation between two or among several states, be regional or encompass states from several regions.

What constitutes 'internationality' may have various interpretations. It is probably safe to say that an arrangement for co-operation in the nuclear field among a few states belonging to the same political or regional environment might not be accepted as creating those conditions for international control which would give adequate assurance of adherence to a world-wide non-proliferation régime. This will, however, depend on such circumstances as the states involved, the conditions of the arrangements, and provisions for further supervision by an outside organization.

A primary condition for internationalization is that it contains provisions designed to trigger an immediate reaction by the other parties if one of them were to attempt to take a unilateral decision to make a proscribed use of the nuclear material or facilities. Such a reaction would have to bring about or restore all international control in the sense described above. Consequently, it would probably demand the involvement of either a number of countries of different political ideologies, a large number of countries, or the physical control by an individual international body.

Ideally, therefore, a truly 'internationalized' nuclear facility would operate under an international régime which (*a*) would ensure that no single national entity could decide upon the use to which the facility is put; (*b*) would see to it that the facility was operated and its products handled in a manner which did not increase the risk of nuclear proliferation; (*c*) would have the power to enforce its decisions and a degree of control which permitted immediate and effective reaction to any non-compliance, and (*d*) would be so constituted as to generate international confidence that its decisions conformed with the precepts of non-proliferation. The facility would also have to be subject to safeguards applied by an international organization other than the operator.

The wish to avoid the risks of nuclear proliferation is not the only argument, and may not even be the decisive one, for the creation of internationalized facilities; many other political factors and technical and economic issues would come into play. These would include the questions whether there was an international demand for the facility concerned, based on solid technical and economic considerations; whether the political will to internationalize a particular activity existed, and whether the financial and technical means to establish or acquire such an installation could be found.

An increasingly important consideration is the availability of a safe and generally acceptable site. In brief, the promotional factors would probably have to outweigh the disadvantages that might accrue from international control arrangements, particularly for those states which might be in a position to establish and run a facility of the type in question by themselves.

Of course, not every nuclear facility lends itself to internationalization, nor is it necessary to impose international control on all nuclear facilities. Whether it is politically desirable for a given fuel cycle activity to be submitted to international control depends in the first place on whether it could materially contribute to nuclear proliferation. From that point of view, the kinds of plant one thinks about in the first place are enrichment plants, reprocessing installations, fabrication plants handling highly enriched uranium or plutonium, and plutonium storage installations. Such facilities not only give a country a potential proliferation capacity, but they might also lend themselves particularly to internationalization.

V. Summary

These considerations have been analysed in many other contexts. It has been the intention of this paper merely to emphasize, in the concept of the internationalization of the fuel cycle, the element of control which was basic to early proposals for multilateral nuclear co-operation. There are grounds for the belief that the political disturbances and risks which may arise from an interruption of the present non-proliferation régime are caused not only by the actually proven possession of nuclear explosives but also by an assumed capacity to make such explosives. International safeguards, however well devised, are essential as a means of creating confidence, but are basically intended only as *ex post facto* means of detection. The likelihood that in the near future (even if, in the long term, nuclear fuel cycles can be introduced that are less proliferation-prone than some of those so far adopted or in the process of being introduced) an increased number of states will possess facilities that may give them the capacity to make nuclear explosives leads to an increasing perception of the need to apply some sort of international and physical control. It is this control, ensuring immediate and effective action in the interest of non-proliferation, which lies at the heart of the concept of internationalization, and which must determine the way in which any scheme for such internationalization is implemented.

References

SIPRI (Stockholm International Peace Research Institute), 1975. *Safeguards Against Nuclear Proliferation* (Almqvist & Wiksell, Stockholm), p.1.

SIPRI (Stockholm International Peace Research Institute), 1979. *Postures for Non-Proliferation* (Taylor & Francis, London), pp.2–3.

Szasz, P.C., 1979. Sanctions and international nuclear controls, *Connecticut Law Review,* 2(3), Spring.

UN (United Nations), 1977 *The United Nations Disarmament Yearbook* (United Nations Publication, New York), vol.2, chap.9, pp.130–39.

Paper 6. A new international consensus in the field of nuclear energy for peaceful purposes

A.J. MEERBURG*

Ministry of Foreign Affairs, Casuaristraat 16, 2511 VB The Hague, The Netherlands

I. Introduction

The need for a common, world-wide approach towards nuclear energy, in general, and towards nuclear trade and the fuel cycle, in particular, seems self-evident. The present situation—characterized by resentment of the Nuclear Suppliers Group and the US Nuclear Non-Proliferation Act and by scepticism with respect to the background of the International Nuclear Fuel Cycle Evaluation (INFCE), as well as by the fact that Yugoslavia has proposed that a UN Conference on nuclear energy be held—indicates that something is basically wrong. Better international relations in the nuclear field are a necessary prerequisite for an effective, long-term non-proliferation policy, if this is possible at all.

However, hopefully, the worst is over. INFCE and related international talks have taught us a lot. Although INFCE has been unable to solve many of the problems, it has made people aware, particularly those in the nuclear field, that non-proliferation is a real and valid issue. It has taught others that technical fixes for building a more proliferation-resistant fuel cycle are not only difficult to achieve but, moreover, only of limited value compared with the political issues involved in proliferation.[1]

This paper enumerates measures in the field of peaceful nuclear energy which could form or could lead to the new international relationship sought for. Some of the measures ask for national decision-making and consequently, for voluntary implementation. Other measures depend on voluntary arrangements among a few countries. Still others can only be

---

* The opinions expressed in this paper are those of the author and do not necessarily reflect the official policy of his professional affiliation.

[1] For example, those who believe that SALT II fulfils the obligations of the nuclear powers under Article VI of the Non-Proliferation Treaty (NPT) are badly mistaken. The same holds for those who believe that a nuclear arms race in Europe is barely relevant for a world non-proliferation policy.

achieved after international negotiations. The various measures have different time-frames and need different forums. It is therefore impossible to indicate *when* the 'consensus' would be reached or *what,* exactly, it would consist of; the achievement of a more stable relationship in the nuclear field is of necessity a dynamic process. Moreover, it is not at all clear that we will achieve our goal.[2] But we must do our utmost!

Measures could be divided, although not rigidly, into those which affect the fuel cycle technically and those of a more institutional nature. The technical measures have a bearing both on non-proliferation and on physical protection,[3] while the institutional measures mainly concern proliferation by states.

II. A technically safer fuel cycle

A number of technical measures for a more proliferation-resistant fuel cycle are discussed below.

1. Although INFCE did not reach this conclusion, there is no particular reason for building fast breeder reactors in the coming 20--30 years except for research purposes. These research reactors need to be internationalized to emphasize that this important potential energy source is being developed for mankind as a whole.[4]

There is no doubt that fuel cycles in which large amounts of plutonium circulate are more dangerous than those without such materials. On the other hand, the fast breeder can perhaps not be avoided as a future source of energy in view of the limited resources of uranium. These resources are, moreover, unevenly spread over the world. However, there is no need to be hasty. There is sufficient uranium for the coming decades; fast breeders are still very costly; present breeding factors are too low to be really interesting, and in the meantime we would be able to devote our efforts towards the development of other energy sources. Some decades' delay could be important from a non-proliferation point of view. International research in the fast breeder field must continue, however, for a sensible choice to be made sometime during the next century.

---

[2] The explosion of a nuclear device by Pakistan, for example, could destroy the prospects for a new consensus, since it could lead to even stiffer export controls than already exist.

[3] In INFCE one was barely allowed to discuss physical protection since this was considered a national question. However, when making national decisions on the fuel cycle, one must take physical protection very seriously into account.

[4] Other important energy developments, such as fusion and solar energy, need comparable treatment.

2. There is no need for the recycle of plutonium in thermal reactors. Since thermal recycle would introduce plutonium in many parts of the fuel cycle, it would be more dangerous, at present, than the fast breeder. Recycle is of marginal importance from an economic or energy point of view.

3. The amount of reprocessing should be adapted to actual needs. Contrary to earlier beliefs, reprocessing does not seem absolutely necessary for adequate waste handling (except for Magnox fuel). It does not seem as if there will be an appreciable problem with storage of spent fuel for some decades. The option could thus be kept open either to reprocess later or to store indefinitely. From a proliferation point of view, it is much better to keep the plutonium in the spent fuel than to extract it.

4. Active exploration for and exploitation of uranium is called for. More uranium would make it possible to delay the introduction of the fast breeder.

5. Uranium enriched above 20 per cent should no longer be used in any type of reactor.

6. Research could continue with respect to other reasonably safe cycles, in particular the denatured uranium--thorium cycle.

7. Co-conversion and co-reprocessing could be further explored, as could other technical measures which would mainly enhance physical protection, for example, the transport of plutonium as mixed oxide.

III. Institutional arrangements

Institutional arrangements which would contribute towards better international relationships in the field of peaceful nuclear energy are enumerated below.

1. The establishment of a truly international plutonium storage régime is needed and is probably the key to harmonizing relations between nuclear suppliers and consumers. A consensus on export requirements does not seem possible without the abolishment of the present requirement of 'prior consent' with respect to reprocessing. Supplier countries which demand this prior consent would probably be willing to give it up only if they could be convinced that an effective international plutonium storage system were being established. However, such a system must be so reliable that it really creates confidence—among suppliers, that plutonium will not be misused, and among consumers, that they will receive plutonium when they need it in accordance with agreed rules for release.

2. After the establishment of a reliable international plutonium storage régime, or when establishment seems certain, a new set of nuclear export requirements must be developed and approved by both suppliers

and consumers. Such requirements could include: (*a*) a commitment by the recipient that no nuclear explosives will be produced, (*b*) the acceptance of full-scope safeguards, (*c*) bringing plutonium, produced from exported materials or equipment, under the international plutonium storage régime (no prior consent on reprocessing), (*d*) the acceptance of adequate levels of physical protection, (*e*) undertakings regarding retransfer, but no prior consent on retransfer if the new importer accepts the same rules, and (*f*) no unilateral changes in export conditions by the supplier.

It is of the greatest importance that agreement is reached on export requirements that can be accepted by as many suppliers and consumers as possible. This can only be achieved by concessions from both sides: suppliers must be assured that there is no misuse of the exported material, equipment and technology, and recipients must be assured of supply and non-interference. Although procedures should be established for (internationally) changing the export rules, these should not be changed by any one party alone.

3. Special rules are needed for the export of sensitive technology, such as enrichment and reprocessing. A multilateral or international set-up for such plants would be helpful. Export rules on these sensitive technologies need to be stricter than those for the export of technology for other parts of the fuel cycle. Establishing only multilateral or international reprocessing and enrichment plants would, *inter alia,* strengthen the application of International Atomic Energy Agency (IAEA) safeguards. For example, with respect to reprocessing plants, which are difficult to safeguard, multilateralization would make diversion by one participating country more difficult. With respect to ultra-centrifuge enrichment plants, which are easy to safeguard but which can be comparatively easily converted, multilateralization would ensure that the plant stayed under safeguards. Co-location of reprocessing and fuel fabrication plants would also be useful from a safeguards and physical protection point of view.

4. Safeguards need to be strengthened, particularly with respect to large reprocessing and mixed-oxide fuel fabrication plants. Due to technical difficulties in safeguarding such plants because the material unaccounted for becomes too large, new safeguards methods need to be developed. These would probably imply more intrusive access to the plants involved. Countries and operators must be willing to accept such safeguards, and this is a significant problem considering the difficulties the IAEA already encounters.[5] It would also be most helpful if the nuclear weapon states would accept full-scope safeguards on their civilian nuclear industry. Not only would this eliminate an unnecessary discriminatory feature of the present application of safeguards, but it would also pave the way for the verification of a useful arms control measure, that is, a cut-off in the production of fissionable materials for weapon purposes.

[5] For example, one is struck by the fact that countries of the European Community and others try to restrict the IAEA inspectors as much as possible.

5. Assured fuel supply systems could be developed. INFCE Working Group 3 has studied this matter and has made various suggestions, such as the setting up of 'safety nets' between users and the possibility of a fuel bank. These rather complicated arrangements need to be further explored.

6. Arrangements for international management of spent fuel could be useful. Particularly for smaller countries, international spent-fuel stores could make the decision not to reprocess much easier.

IV. Conclusions

The enumeration of possible elements for a new international consensus in the nuclear field is given in this paper only to get an overall picture of possible and useful developments. How these measures could be achieved needs further consideration; in particular, one must examine the role which the IAEA, the second NPT Review Conference and/or other bodies and conferences could play in this process. Some of the measures are already being developed in, for example, the IAEA expert groups on international plutonium storage and on international management of spent fuel. It is our common task to develop these measures further as a contribution to an effective and harmonious non-proliferation régime, which is already encountering great political difficulties in many other respects.

Paper 7. Some factors affecting prospects for internationalization of the nuclear fuel cycle

W.H. DONNELLY*

Congressional Research Service, Library of Congress, Washington, DC 20540, USA

I. Introduction

Since concern over nuclear weapon proliferation revived in the mid-1970s, the idea that its risks can be decreased by placing the sensitive parts of the nuclear fuel cycle under international management or control has become a kind of conventional wisdom that can allay the fears about the expanding use of nuclear energy increasing risks of nuclear conflagration. Clearly, the importance attached to preventing the further spread of nuclear weapons derives from the fear that the United States and the Soviet Union might blunder into nuclear Armageddon because nuclear weapons might be used by some other nation or perhaps even a terrorist or extremist organization.

II. Foundations of US policy on internationalization

The foundation for US policy on internationalization of the nuclear fuel cycle was built fragmentally from 1945–46, when the United States advanced the Baruch Plan in the United Nations Atomic Energy Commission and when Congress enacted the Atomic Energy Act of 1946, until President Carter's term and the Nuclear Non-Proliferation Act of 1978. During the intervening years, a few new building blocks were added in the form of the Atomic Energy Act of 1954. Today there is a confluence of legislation and Presidential policy supportive of several approaches to international operations and control for sensitive parts of the fuel cycle, including spent fuel storage and enrichment, but much less supportive of reprocessing.

---

* The opinions expressed in this paper are those of the author and do not necessarily reflect the official policy of his professional affiliation.

The Atomic Energy Act of 1954

The combination of Atoms for Peace and a burst of optimistic enthusiasm from the infant nuclear energy industry in the early 1950s culminated in a major overhaul of US nuclear legislation in the Atomic Energy Act of 1954. Whereas the 1946 Act had been drafted in an effort to keep the secret of the atom bomb, the 1954 Act encouraged the dissemination of scientific and technical information. Concerning international co-operation, the Act provided for bilateral agreements for nuclear co-operation with nations or regional defence organizations and also authorized the President to enter into an international arrangement with a group of nations in international co-operation in the non-military application of atomic energy.

President Ford's policy

On 28 October 1976, Presidential Candidate Ford made a major and comprehensive statement on nuclear policy. Many of the initiatives, policy decisions and actions he announced related to the goals and functions of international organization to control proliferation. Those most directly related included the following:

(*a*) A policy decision that reprocessing and recycling of plutonium should not proceed unless there was sound reason to conclude that the world community could "effectively overcome the associated risks of proliferation". However, the USA and other nations could and should increase their use of civil nuclear power even if reprocessing and recycling were found to be unacceptable. Likewise, the United States should greatly accelerate its diplomatic initiatives to control the spread of plutonium and technologies for separating plutonium.

> Effective nonproliferation measures will require the participation and support of nuclear suppliers and consumers. There must be coordination in restraints so that an effective nonproliferation system is achieved, and there must be cooperation in assuring reliable fuel supplies so that peaceful energy needs are met. (Ford, 1976: 1626)

(*b*) An invitation to all nations to join with the United States in exercising maximum restraint in the transfer of reprocessing and enrichment technology and facilities by avoiding such sensitive exports or commitments for at least three years.

Ford urged nuclear supplier nations to provide nuclear consumers with fuel services, instead of sensitive technology or facilities, and to continue study of the idea of a few suitably sited multinational fuel cycle centres to serve regional needs, when effective safeguards and economy warranted.

To these ends, he directed Secretary of State Kissinger to initiate consultations with other nations on arrangements to co-ordinate fuel services and to develop other means of ensuring that suppliers would be able to offer, and consumers would be able to receive, an uninterrupted and economical supply of low-enriched uranium fuel and fuel services. Additionally, Kissinger was to negotiate for disposition of spent fuel with consumer nations that adopted responsible restraints.

Concerning international control against proliferation, Kissinger was to pursue "vigorous discussions" to establish a new international régime for storage of civil plutonium and spent reactor fuel. The United States, he recalled, had made such a proposal to the IAEA in the spring of 1976. Of this he said:

> Creation of such a regime will greatly strengthen world confidence that the growing accumulation of excess plutonium and spent fuel can be stored safely, pending reentry into the nuclear fuel cycle or other safe disposition. I urge the IAEA, which is empowered to establish plutonium depositories, to give prompt implementation to this concept.
>
> Once a broadly repesentative IAEA storage régime is in operation, we are prepared to place our own excess civil plutonium and spent fuel under its control. Moreover, we are prepared to consider providing a site for international storage under IAEA auspices. (Ford, 1976: 1628)

President Carter's policies

Several of Carter's statements, as a candidate and as president, touched upon internationalization. As a presidential candidate, he spoke about this subject on two occasions. On 13 May 1976 he addressed a UN conference where he warned about the hazards of proliferation and about the limitations of IAEA safeguards. Nuclear energy, he said, must be at the very top of the global challenges that call for new forms of international action. Concerning enrichment and reprocessing, he said in part:

> We should also give the most serious consideration to the establishment of centralized multinational enrichment facilities involving developing countries' investment participation, in order to provide the assured supply of enriched uranium. And, if one day, as their nuclear programs economically justify use of plutonium as a supplementary fuel, similarly centralized multinational reprocessing services could equally provide for an assured supply of mixed oxide fuel elements, (Carter, 1976)

Spent fuel storage was also an option for international action. Of this, he said:

> One final observation in this area: the need to cut through the indecision and debate about the long-term storage of radioactive wastes and start doing something about it...Waste disposal is a matter on which sound international arrangements will clearly be necessary. (Carter, 1976)

As the campaign drew near its close, Carter said, in an address on 25 September 1976 at San Diego, that as president he would take eleven specific steps to control proliferation. One would be to explore international initiatives, such as multinational enrichment plants and multinational spent fuel storage, as alternatives to the establishment of enrichment or reprocessing plants on a national basis.

A few months after his inauguration, President Carter addressed nuclear power policy in a major statement on 17 April 1977. Voicing deep concern about the consequences of the uncontrolled spread of nuclear weapon capability, he said, "We can't arrest immediately and unilaterally. We have no authority over other countries. But we believe that these risks would be vastly increased by the further spread of reprocessing capabilities..." He announced seven steps of which four touched upon one or another aspect of international nuclear relations. These included: (*a*) an increase in US enrichment capacity to provide adequate and timely supplies to countries that need them so that they will not be required or encouraged to reprocess their own materials, (*b*) proposal of legislation to permit the United States to sign supply contracts and remove the pressure for the reprocessing of nuclear fuels by other countries, (*c*) continuation of the US embargo of export of either equipment or technology for enrichment or reprocessing, and (*d*) continuation of discussions with supplier and recipient nations of a wide range of international approaches and frameworks that will permit all countries to achieve their own energy needs while at the same time reducing the spread of the capability for nuclear explosive development. Elaborating on this, President Carter said:

> Among other things—and we have discussed this with 15 or 20 national leaders already—we will explore the establishment of an international nuclear fuel cycle evaluation program so that we can share with countries that have to reprocess nuclear fuel the responsibility for curtailing the ability for the development of nuclear explosives.
>
> One other point that ought to be made in the international negotiation field is that we have to help provide some means for the storage of spent nuclear fuel materials which are highly explosive, highly radioactive in nature. (Carter, 1977)

The Nuclear Non-Proliferation Act of 1978

On 10 March 1978 President Carter approved the Nuclear Non-Proliferation Act, culminating a sustained Congressional initiative, later joined by the Administration, to specify US non-proliferation policies. Significant parts of the Act which relate to internationalization include the statement of policy and provisions for US initiatives to provide adequate nuclear fuel supplies, authority for the International Nuclear Fuel Cycle Evaluation, US negotiations for principles and procedures in case of violations of non-proliferation commitments, and steps to seek agreement on certain export policies.

In its statement of policy, Congress declared that the proliferation of nuclear explosive devices or of the direct capability to manufacture or otherwise acquire such devices poses a grave threat to the security interests of the United States and to continued international progress towards world peace and development. Accordingly, it is the policy of the United States to:

> actively pursue through international initiatives mechanisms for fuel supply assurances and the establishment of more effective international controls over the transfer and use of nuclear materials and equipment and nuclear technology for peaceful purposes in order to prevent proliferation, including the establishment of common international sanction. (NNPA, 1978*a*)

Concerning nuclear fuel supply, the Act directed the President to initiate prompt discussions with other nations and groups of nations to develop international approaches for meeting future world-wide nuclear fuel needs. In particular, he is to negotiate with nations possessing nuclear fuel production facilities or source materials, and other nations and groups of nations, such as the IAEA, as appropriate, for the timely establishment of binding international undertakings for the following six purposes:

(1) Establishment of an international fuel authority to provide fuel services and to allocate fuel resources.

(2) Establishment of conditions for the authority's fuel supply assurances. The President is to seek to ensure that its benefits are available to non-nuclear weapon states only if they accept IAEA safeguards on all their peaceful nuclear activities, do not manufacture or otherwise acquire nuclear explosives, do not establish any new enrichment or reprocessing facilities, and place existing facilities under "effective international auspices and inspection".

(3) Determination of feasible and environmentally sound approaches for effective international control of siting, development, management and inspection of nuclear fuel service facilities, including those for the storage of special nuclear material.

(4) Establishment of repositories for the storage of spent nuclear reactor fuel under effective international auspices and inspection.

(5) Arrangement for nations that place spent fuel in such repositories to receive appropriate compensation for the energy content of the spent fuel if recovery of such energy content is deemed necessary or desirable.

(6) Establishment of sanctions for violation or abrogation of these undertakings.

Concerning the International Nuclear Fuel Cycle Evaluation (INFCE), initiated in October 1977 by President Carter, the Act directed the President to invite all nuclear supplier and recipient nations to re-evaluate all aspects of the nuclear fuel cycle, emphasizing alternatives to an economy based on separation of pure plutonium or the presence of highly enriched uranium, methods to deal with spent fuel storage, and methods to improve safeguards for existing nuclear technology.

Concerning countermeasures for violation of non-proliferation commitments, the Act provides for the United States to seek to negotiate with other nations and groups of nations to:

(1) adopt general principles and procedures, including common international sanctions, to be followed in the event that a nation violates any material obligation with respect to the peaceful use of nuclear materials and equipment or nuclear technology, or in the event that any nation violates the principles of the Treaty, including the detonation by a non-nuclear weapon state of a nuclear explosive device; and

(2) establish international procedures to be followed in the event of diversion, theft, or sabotage of nuclear materials or sabotage of nuclear facilities, and for recovering nuclear materials that have been lost or stolen, or obtained or used by a nation or by any person or group in contravention of the principles of the Treaty. (NNPA, 1978*b*)

Concerning international nuclear export policies, in addition to directing the President to seek agreement on five specified conditions for transfers—conditions closely resembling those prescribed for approval of US exports—in negotiation or renegotiation of agreements for nuclear co-operation, the Act again emphasizes the idea that certain nuclear activities should be under international auspices and inspection. Here the specified goal is international agreement that no source or special nuclear material within the territory of any nation or group of nations under its jurisdiction or under its control anywhere will be enriched or reprocessed, no irradiated fuel elements will be altered in form or content, and no fabrication or stockpiling involving plutonium, uranium-233, or uranium enriched to more than 20 per cent shall be performed except in a facility under effective international auspices and inspection, and any irradiated fuel elements shall be transferred to such a facility after removal from a reactor, as soon as practicable consistent with safety requirements. Such facilities should be limited in number and carefully sited and managed so as to minimize proliferation and environmental risks. In addition, non-nuclear weapon states other than the host country should be limited in their access to sensitive nuclear technology associated with such facilities. Also, any facilities for the short-term storage of fuel elements containing plutonium, uranium-233, or uranium enriched to more than 20 per cent, or of irradiated fuel elements before their transfer to a storage facility, should be under effective international auspices and inspection.

The Act does not define "effective international auspices and inspection". The Senate report which brought the Act to the floor explained that there was no definition because considerable study and negotiation would be required to determine precisely what arrangements would be most desirable from the standpoint of reducing proliferation risks, ensuring economic viability, and achieving widespread participation. However, it was expected that as a minimum there should be comprehensive IAEA safeguards, including continuous inspection, if such measures are deemed necessary by the IAEA to ensure early detection of any diversion of material. In addition, there should be international or multinational participation in the operation and/or ownership of the facility, preferably including at least one nuclear weapon state; international or multinational ownership of the fuel; a commitment to impose sanctions should any non-nuclear weapon state or subnational group seek to acquire unauthorized access to the facility or to materials located at it; and an enforceable ban on the establishment of national sensitive fuel cycle facilities in nations which are able to obtain adequate fuel services from the international facilities (US Senate, 1977).

III. The International Nuclear Fuel Cycle Evaluation

Since other papers have amply addressed INFCE, it is sufficient here to note that its outcome is likely to affect prospects for internationalization of the nuclear fuel cycle. INFCE can be expected to support further internationalization, such as is proposed in the US Nuclear Non-Proliferation Act, particularly because of the INFCE consensus that institutional as well as technological measures are required to restrain the proliferation risks of nuclear power.

IV. The second NPT Review Conference

Another factor likely to affect the prospects for internationalization is what happens at the second Review Conference of the Non-Proliferation Treaty now scheduled for summer 1980. If the 1975 Review Conference is an indicator, many non-nuclear weapon states are again likely to raise complaints of discrimination against them, both in the NPT régime itself and in export policies of the United States and the nuclear exporter countries, and to revive charges of nuclear imperialism. Also, considering the recent UN Conference on Science and Technology, developing countries at the Review Conference may press for more rather than less assistance with the various technologies of the nuclear fuel cycle. Such trends would probably be unfavourable to proposals for new ventures in internationalization, particularly if the developing countries attach more importance to their possession of the paraphernalia of nuclear power than to enjoyment of its benefits assured by reliable external supply.

After INFCE and the NPT Review Conference, it seems unlikely that any substantial negotiations for internationalization can occur before next autumn. In that case, the attitudes and positions taken by nations at the Review Conference can reasonably be expected to signal their future negotiating positions. This suggested linkage of the Conference with the success of internationalization adds to its importance and calls attention to the importance of the purpose, status and co-ordination of plans by the leading nuclear power countries for their participation in it.

V. Other factors affecting prospects for internationalization

Many other factors can increase, reduce or divert the impetus for internationalization. Well recognized and needing no further discussion are the dangers of proliferation, the practicability of denial, and the incentives of an assured supply of nuclear fuels and services. Other factors appear less frequently in debates on proliferation, but merit attention. Several are briefly mentioned below, including the world's energy and economic outlooks, the future of nuclear power, and the state of US influence in nuclear matters.

Outlook for the world's economy and energy

The inescapable message of the Arab oil embargo of 1973–74 is that the outlook for the world's economy is linked to the adequacy, reliability and price of its energy supplies. In a world of anxiety and dependence with regard to energy resources, it will be increasingly difficult to finance the capital and operating costs of internationalization, and countries are likely to hold that what funds are available must go for urgent, visible problems of today rather than for perhaps theoretical problems of tomorrow.

The future of nuclear power

The reason for attention to internationalization is the link between nuclear power and weapon proliferation. The more nuclear power spreads, the greater is the perceived risk. World forecasts of nuclear power have been dropping with the drop in projections for electricity caused by a sluggish world economy. As the costs of nuclear power plants increase, fewer countries can afford them. Also, opposition to nuclear power continues unabated in many countries. The accident at the Three Mile Island Nuclear Power Plant in the United States in March 1979 has fanned the fears and efforts of many opponents in many countries. Controversy over the long-term disposition of spent fuel and nuclear waste has produced restrictive government actions, most notable in FR Germany and in Sweden. The early 1980s are likely to be a critical time for nuclear power, as leading supplier countries try to deal with their capacities to produce more nuclear power plants than the world market can absorb.

The state of US influence

When the NPT was signed, the United States was the principal supplier of enriched uranium for the Western World; its manufacturers dominated the Western world market for nuclear power equipment, and its trading partners saw it as a reliable supplier. Now, some expect the Soviet Union to deliver more enrichment to Europe than the United States will. Eurodif and Urenco are hungry for orders, while prices of US enrichment have soared, and the reliability of its supply under new US policies is not yet fully re-established. So US nuclear influence in 1980 may be weaker, but it is by no means over. Lacking its former monopoly, the United States will have to look more to the force of its ideas and to the search for common interests.

VI. Conclusions

For the moment, internationalization promises to provide reason to go ahead, keeping in mind that virtue unaided may not triumph. Many factors will affect the ultimate outcome. Some will be susceptible to direction and change, while others will not. The outcome of both INFCE and the coming NPT Review Conference, as well as future prospects for nuclear energy, are two such factors. Every relevant factor deserves continued attention if we are to avoid surprises and the temptation to assume that the expected non-proliferation benefits of internationalization are, of themselves, sufficient to assure its success, especially as internationalization moves from the offices of analysts to the conference rooms of diplomats and political leaders.

References

Carter, J., 1976. Nuclear energy and world order, address at the United Nations, 13 May.

Carter, J., 1977. Nuclear power policy, the President's remarks announcing his decisions following a review of US policy, April 7, 1977, *Weekly Compilation of Presidential Documents,* Vol. 13, 11 April, p.504.

Ford, G., 1976. Nuclear policy, Statement by the President, October 28, 1976, *Weekly Compilation of Presidential Documents,* Vol.12, 1 November.

NNPA (Nuclear Non-Proliferation Act), 1978*a*. Section 2, *US Statutes*. Vol. 92, p.120.

NNPA (Nuclear Non-Proliferation Act), 1978*b*. Section 203, *US Statutes*, Vol. 92, p.124.

US Senate, 1977. Nuclear Non-Proliferation Act of 1977 Report, together with additional views to accompany S.897, Senate Report 95-467, 3 October, p. 25.

Paper 8. International plutonium policies: a non-proliferation framework

D.W. CAMPBELL and M.J. MOHER*

Commodity and Energy Policy Division, Department of External Affairs, Ottawa, Canada

I. Introduction

This paper outlines a non-proliferation framework which could govern the legitimate use of plutonium and reprocessing technology by harmonizing and building upon relevant elements of the current non-proliferation régime. It takes as its starting-point that some reprocessing of irradiated fuel will take place for a variety of reasons, that separated plutonium will therefore be available in increasing quantities, and that this separated plutonium will sooner or later be used by a number of countries in their nuclear fuel cycles. While considerations other than non-proliferation are also taken into account, a procedure and possible criteria to be used within that procedure which would minimize the proliferation risk associated with reprocessing and separated plutonium are identified. The basic assumption is that without an international consensus on this point the significant long-term political, financial and technological commitments necessary for the economical development and use of specific technologies in the nuclear fuel cycle or for the internationalization of specific elements of that cycle may be hampered or foreclosed.

II. Elements of the non-proliferation régime

The current non-proliferation régime is composed of international treaties, multilateral agreements and national policies. In fact it is undoubtedly misleading to speak of *the* current régime since there is no single inter-

*The opinions expressed in this paper are those of the authors and do not necessarily reflect the official policy of their professional affiliations.

nationally agreed set of non-proliferation measures accepted by all countries with significant nuclear power programmes. The difficulties arising from this situation have been addressed by Working Group 3 of the International Nuclear Fuel Cycle Evaluation (INFCE), which has concluded that "a reduction or mitigation of uncertainties in the exercise of [non-proliferation] export and import controls would assist both suppliers and consumers to protect their mutual interest in stable nuclear trade consistent with non-proliferation and energy needs". It is generally agreed that this task should receive priority attention in the post-INFCE period.

In approaching this task, it is worthwhile to identify the elements of the current non-proliferation régime which bear on reprocessing and plutonium use. These elements can be found in: (*a*) the Statute of the International Atomic Energy Agency (IAEA) and the Agency's safeguards operations; (*b*) the Non-Proliferation Treaty (as well as regional treaties such as the Euratom Treaty and the Treaty of Tlatelolco); (*c*) the guidelines of the Nuclear Suppliers Group (NSG), and (*d*) national nuclear export policies.

When the IAEA Statute was being drafted over 20 years ago, the reprocessing of irradiated fuel and the use of the resulting separated plutonium were regarded as legitimate aspects of the nuclear fuel cycle, particularly that based on light water reactors (LWRs), but was considered to be of such proliferation significance that the Agency was given certain rights and responsibilities under Article XII.A.5 of its Statute:

> To approve the means to be used for the chemical processing of irradiated materials solely to ensure that this chemical processing will not lend itself to diversion of materials for military purposes and will comply with applicable health and safety standards; to require that special fissionable materials recovered or produced as a by-product be used for peaceful purposes under continuing Agency safeguards for research or in reactors, existing or under construction, specified by the member or members concerned; and to require deposit with the Agency of any excess of any special fissionable materials recovered or produced as a by-product over what is needed for the above-stated uses in order to prevent stockpiling of these materials, provided that thereafter at the request of the member or members concerned special fissionable materials so deposited with the Agency shall be returned promptly to the member or members concerned for use under the same provisions as stated above.

The NPT did not single out plutonium as such for particular attention, but under Article III non-nuclear weapon states party to the Treaty undertake:

> to ensure that safeguards will be applied to all special fissionable material in all peaceful nuclear activities within the territory of the state, under its jurisdiction or carried out under its control anywhere;

while states party undertake:

> not to export special fissionable material or any equipment or material especially designed or prepared for the processing, use or production of special fissionable material to any non-nuclear-weapon State for peaceful purposes unless the special fissionable material shall be subject to the safeguards required by this article.

The Nuclear Suppliers Group (NSG) gave special attention to plutonium and to the technology and equipment for obtaining it in the formulation of its guidelines (INFCIRC 254). Under those guidelines, members of the NSG agreed that the transfer of reprocessing technology or specially designed equipment for reprocessing should take place only when covered by IAEA safeguards with duration and coverage provisions in conformance with GOV/1921 guidelines. Any such transfers were to require an undertaking that IAEA safeguards would apply to any facility of the same type constructed during an agreed period in the recipient country. Moreover, members of the NSG defined reprocessing technology as 'sensitive' and agreed to exercise restraint in the transfer of sensitive facilities, technology and weapons-usable materials. If such transfers were to take place, members undertook to encourage recipients to accept, as an alternative to national plants, supplier involvement and/or other appropriate multinational participation in the resulting facilities. Members also agreed to promote international (including IAEA) activities concerned with multinational regional fuel cycle centres.

Equally significant as regards plutonium itself, members of the NSG recognized the importance of including in agreements on the supply of nuclear materials or of facilities which produce weapons-usable material, whenever appropriate and practicable, provisions calling for mutual agreement between the supplier and the recipient on arrangements for reprocessing, storage, alteration, use, transfer or retransfer of any weapons-usable material involved.

The evolution of the international attention given to the risk of proliferation arising from reprocessing and the increasing quantities of plutonium available in the nuclear power fuel cycle was also reflected in the nuclear export policies of a number of countries. In general and perhaps oversimplified terms, the approaches of these countries to reprocessing and plutonium use effectively fall into three categories. The first approach was based on the view that the use of plutonium in current nuclear fuel cycles is uneconomical and unnecessary, that the proliferation risks are of major proportions, and therefore that reprocessing—and hence the production and use in the fuel cycle of separated plutonium—can and should be deferred indefinitely. The second approach differed in that it did not necessarily seek the deferral of reprocessing, but tended to recognize that reprocessing and plutonium use in the fuel cycle will in some cases take place. This approach therefore focused more on means to minimize the risks of proliferation arising from these activities. The third approach, while not denying that there are proliferation risks associated with reprocessing and plutonium use, was adopted by those countries who believed that these proliferation risks could be effectively minimized and that reprocessing and plutonium use were essential to their national nuclear energy programmes.

While all three approaches possess a common element in that the proliferation risks associated with reprocessing and plutonium use are recognized, it is obvious that the various assessments of the extent of those risks vary considerably. In this context it is significant to note that several countries, who are among themselves major suppliers of uranium and of enrichment services, have established as an element of their national policies a requirement that countries with which they are engaged in nuclear co-operation must provide them with a prior-consent right over the reprocessing of nuclear material subject to their respective intergovernmental agreements and over the subsequent use of the separated plutonium.

The latter point is worth stressing. In the case of Canada, for example, considerable importance is attached to this prior-consent right. The objective of Canada in seeking recognition of this right from its nuclear partners is purely and solely to promote the non-proliferation objective; that is, given that some reprocessing is going to take place, steps must be taken to ensure that reprocessing will be carried out in such a way that the risk of proliferation is minimized. As a major exporter of uranium for use as a nuclear fuel in both heavy water reactors (HWRs) and LWRs and as an exporter of nuclear reactors and other components of the nuclear fuel cycle, Canada believes that it has a major responsibility to ensure that this is so. The political ability of the Canadian government to sustain its nuclear exports is, given the commitment of the Canadian people to non-proliferation, dependent on the maintenance of public and parliamentary confidence in the adequacy of Canada's non-proliferation arrangements. Reprocessing and plutonium use are seen as particular areas of concern in this context.

At the same time, the legitimate concern of a country, having provided this right of prior consent, that its nuclear partner might act arbitrarily in the future and thereby have an adverse impact upon its nuclear power programme must be recognized. Working Group 3 of INFCE agreed that "where the right of prior consent exists the criteria for the exercise of that right should be established, to the extent possible, before long-term fuel supply contracts are concluded or, for short-term contracts, before fuel is committed to power reactors". It was felt that the establishment of such criteria would increase the predictability of the exercise of the right of prior consent and thus facilitate longer-term nuclear fuel cycle programme planning.

III. A non-proliferation framework

In the formulation of a non-proliferation framework applicable to reprocessing and plutonium use and designed to facilitate international co-operation with regard to those activities, there are therefore three fundamental elements upon which to meditate: (*a*) Article XII.A.5 of the

IAEA Statute, (*b*) the undertakings of the NSG in this context, and (*c*) the right of prior consent over reprocessing and the subsequent use and storage of separated plutonium required by some countries. Not all three of these elements of course impinge on the nuclear fuel cycle at the same time and place.

The latter observation recognizes a key factor which should be recognized from the beginning. A nuclear fuel cycle incorporating reprocessing and plutonium will in most cases involve several countries over a substantial period of time. This can be illustrated as follows:

1. A contract would be signed for uranium which may not be delivered for a period of 10 years or longer.
2. Before its use in an LWR, that uranium would have to be enriched and then fabricated into fuel elements.
3. The fuel would next be placed in a power reactor, perhaps after some time in a national or user stockpile.
4. The irradiated fuel would then have to be allowed to cool for a period of time.
5. The cooled irradiated fuel would then have to be transferred to a reprocessing facility.
6. The irradiated fuel would be reprocessed and the separated plutonium stored for subsequent use.
7. The separated plutonium would be used.

This general sequence is not meant to be definitive in any way but to illustrate that more than one or two countries would likely acquire a non-proliferation interest with regard to the nuclear material in question and that this interest would most likely have to be demonstrated over a period of time possibly reaching 20–30 years into the future.

One could of course assess each step in the above sequence. However, two stages will be identified and examined in this paper: the first stage will include the pre-reprocessing process while the second will include the post-reprocessing or plutonium storage and/or use process.

The pre-reprocessing stage

There are two basic scenarios which should be considered insofar as this stage is concerned. These scenarios are developed around the fact that reprocessing can take place either in national facilities accepting national and/or foreign irradiated fuel, *or* in multinational facilities which accept irradiated fuel from those countries who co-operated in the establishment of the facilities as well as, possibly, from elsewhere (see figure 1). It is evident from these scenarios that two or more countries will likely have an interest in where this reprocessing will take place, whether that interest arises in the exercise of a prior-consent right over reprocessing or as a reflection of a more general non-proliferation commitment.

Figure 1. Two pre-reprocessing scenarios

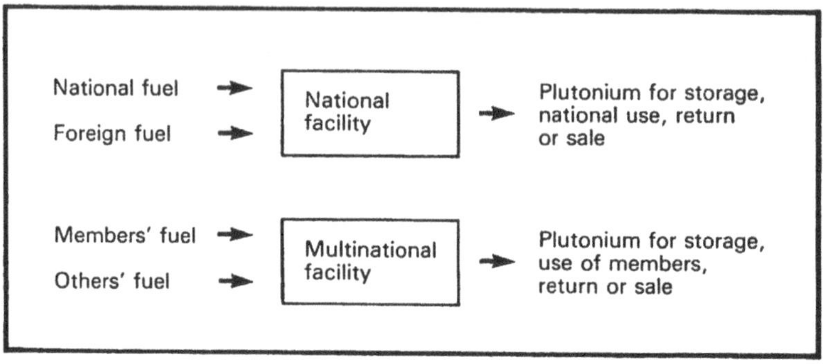

From the non-proliferation perspective it is obviously essential that the reprocessing facility be in a country whose non-proliferation credentials are without fault. It is therefore suggested that one *sine qua non* should be that reprocessing should take place only in nuclear weapon states that have subjected their reprocessing facilities to international safeguards or in non-nuclear weapon states that have made a binding commitment to non-proliferation (i.e., no explosive use) and accepted international safeguards on all of their nuclear activities. In certain specific circumstances, that is, where geopolitical conditions are such that the long-term non-proliferation commitment of the country concerned may come under unusual stress, it may be necessary to consider whether reprocessing should in fact take place in that country.

The next questions which would have to be addressed concern the reason or reasons for the reprocessing and the quantity to be reprocessed. The interrelationship between these two is clear. Reprocessing should not be carried out merely as an end in itself. Thus any reprocessing activity should form part of a carefully defined programme, and the quantities to be reprocessed should be justified by such programmes. The point of raising this question is not to say 'no to reprocessing' but to say 'no to unnecessary and unjustified reprocessing'. This is not meant to imply that one government should necessarily have the right to approve another government's programme, but is directed towards ensuring that a valid programme exists and that the activities carried out within that programme are 'transparent' from the non-proliferation viewpoint. There should therefore be a readily demonstrated relationship between the reason for which reprocessing is to be carried out and the quantity to be reprocessed.

Thirdly, considerable attention should be paid to how the reprocessing is to be carried out. The reprocessing facility should obviously be one which can be effectively safeguarded, since it is in this sensitive activity that plutonium will exist in a relatively pure form. To date, experience in safeguarding such facilities is limited. While this experience has been useful for the evolution of new and improved safeguards techniques, it is generally recognized that for future reprocessing plants it will be essential to take full account of the criteria for effective safeguards from the inception of plant design and to consider carefully the resultant costs.

In any formulation of an international non-proliferation framework, these questions—the where, why, how and how much of reprocessing—should be answered satisfactorily if a generally acceptable, effective non-proliferation régime that will provide the framework for international co-operation or for the internationalization of specific fuel cycle activities is to be devised.

The post-reprocessing stage

As the result of the first stage, a quantity of plutonium will have been separated and be 'sitting at the back door' of the reprocessing facility awaiting storage or use. It is suggested that an approach similar to that already employed could be used again. No matter what type of reprocessing facility (i.e., national or multinational) is involved, plutonium separated at that facility will be available (*a*) for storage, (*b*) for national use, (*c*) for return to the supplier of irradiated fuel, or (*d*) for sale to another party. Similar questions can therefore be posed:

(*a*) Where should the plutonium be stored?

(*b*) For what use (why) and therefore in what quantity (how much)? Should the plutonium be used either in the country where the reprocessing took place *or* in the country-of-origin of the irradiated fuel *or* in a third country?

(*c*) How should the plutonium be stored or transferred (i.e., in what form and under what conditions)? Once again these questions should be effectively answered if an agreed non-proliferation framework is to be devised.

The latter point is particularly interesting in view of some of the information presented earlier. When reviewing the NSG guidelines, it was noted that members of that group have agreed to endeavour to include, whenever appropriate and practicable, provisions in agreements on the supply of nuclear materials or of facilities that produce weapons-usable material calling for mutual agreement between the supplier and the recipient on arrangements for reprocessing, storage, alteration, use, transfer or retransfer of any weapons-usable material involved. Moreover, we have observed that most of the countries currently involved in reprocessing or planning to be so involved are members of this group. Therefore, most countries that are likely to be involved in reprocessing generally recognize that there should be mutual agreement (or, in other words, prior consent by the supplier) between the supplier and recipient with regard to the storage, alteration, use, transfer or retransfer of any separated plutonium. Thus the questions posed above obviously are relevant.

With regard to the first question posed above, there are advantages in having the plutonium stored at the reprocessing facility since this could reduce the possibility of diversion or of theft. The International Plutonium Storage (IPS) concept, coupled where appropriate with the idea of co-location of facilities, is of considerable merit in this regard. While these concepts are discussed and defined in detail in other papers published in this volume, it is suggested that an element of any international non-proliferation framework should be the requirement that any separated plutonium must be stored under the auspices of an effective IPS scheme.

The second question above is directly related to the same question posed with regard to reprocessing itself. Any supplier with a non-proliferation hold over irradiated fuel and special fissionable material contained in that fuel, as well as any supplier of reprocessing services, will have legitimate concerns with regard to the use for which the separated plutonium is required as well as the quantity of plutonium requested. The answers to both of these can fall into three general categories: (*a*) for research in nuclear physics, and/or (*b*) for research into nuclear fuel cycle technologies, and/or (*c*) for use in thermal recycle and/or fast breeder reactors (FBRs). These answers would presumably have to be evaluated by the original supplier of the nuclear material in question, if appropriate, and against the arrangement of the IPS scheme before any plutonium was transferred from the storage centre.

It is difficult to determine theoretically the form in which plutonium should be stored or subsequently transferred to another facility. The answers will be subject to operational decisions of the reprocessor, to environmental and safety concerns, to physical security considerations and to the end-use for the material concerned. As a general rule, however, it is suggested that the plutonium should be stored and transferred in the most proliferation-resistant form possible in view of prevailing circumstances. Effective safeguards techniques with regard to the facility (i.e., fuel fabrication, reactor, etc.) in which the plutonium is to be used as well as physical protection measures should be developed and employed. With regard to the international transfer of separated plutonium, the successful conclusion of current discussions of a convention covering the international transportation of nuclear material and acceptance of that convention by the states concerned would be highly valuable.

IV. Conclusions

This paper has covered the various elements of the current non-proliferation régime with regard to reprocessing and plutonium use. It is evident that if reprocessing and plutonium storage and use are to take place in a secure commercial environment that recognizes and responds to the non-proliferation concerns of all the countries involved, a consensus on an effective international non-proliferation framework covering these fuel cycle activities will be essential. Moreover, this consensus is a prerequisite for the internationalization of certain nuclear fuel cycle activities.

Essentially this paper suggests that internationally agreed criteria, by which suppliers' prior-consent rights as well as the non-proliferation concerns of reprocessors will be satisfied, should be defined. This step would hopefully go a long way towards meeting the legitimate industrial/commercial concerns of those engaged or intending to be engaged in reprocessing or in using plutonium in their nuclear fuel cycle programmes while recognizing the equally legitimate proliferation concerns of suppliers of nuclear material, equipment, technology or services. It is also suggested that these criteria should cover the 'where', the 'why', the 'how', and the 'how much' of reprocessing and plutonium use. The development of these criteria is seen as being a priority task for the international community concerned. In this regard, it is worth noting that INFCE Working Group 3 agreed:

> that, to meet the concerns of some consumer countries about differences in some of the non-proliferation conditions of bilateral agreements, common approaches would need to be sought against the background of the need to make nuclear power available to all nations who wish to use it for peaceful purposes and the need to achieve this in a way that avoids proliferation while respecting the sovereignty of nations and the national needs of technological development.

With this in mind, the international non-proliferation framework with regard to reprocessing and plutonium use might be based on a consensus among suppliers, reprocessors and other interested states on the following:

(*a*) a recognition that the reprocessing of irradiated fuel and the resulting separated plutonium pose a proliferation risk which merits specific measures to minimize that risk;

(*b*) while respecting the sovereignty of nations and the national needs of technological development, a recognition that reprocessing should take place only when, where and to the extent justified by national or multinational programmes;

(*c*) a commitment by the countries concerned to ensure that international safeguards can be effectively applied to reprocessing facilities, to plutonium stores and to other facilities where plutonium is used;

(*d*) a commitment to apply adequate physical protection measures at all times, and

(*e*) a recognition that, pursuant to (*b*) above, criteria should be developed by the international community concerning the 'where', 'why', 'how' and 'how much' of reprocessing and plutonium use.

An international non-proliferation framework incorporating these five points and employing, where possible, the concepts of international plutonium storage, international spent-fuel storage, multinational fuel cycle centres, and international nuclear fuel banks would minimize the proliferation risks associated with reprocessing and the resultant stocks of separated plutonium, while providing a context within which industrial/commercial activities on the national or multinational level might proceed with a considerable degree of certainty. It is suggested that it is this task which now confronts the international nuclear energy community.

Paper 9. Export of nuclear materials

R.W. FOX*

Australian High Commission, Australia House, Strand, London WC2B 4LA, UK

I. Introduction

Nuclear trade suffers an impediment not commonly found in other forms of trade, that of lack of public acceptance. This can result in a supply–demand stricture, creating problems and uncertainties at both ends. Therefore it is in the interests of international trade that public confidence be bolstered in relation to (*a*) peaceful use; (*b*) safety of operation; (*c*) disposal of wastes, and (*d*) physical protection of plants and materials.

Undoubtedly, a supplier's responsibility is to be sure that what it supplies is not contributing to the risk of nuclear war. But assurances of use for peaceful purposes should not be regarded simply as requirements imposed by suppliers. There can be solid advantages for a consumer, from a trade point of view, if it can establish that a particular acquisition will be used for peaceful purposes only.

It is patently the fact that existing international and bilateral agreements have not created a sufficient degree of confidence that there will be no diversion of weapons-usable material. The Non-Proliferation Treaty (NPT) is a valuable measure, which should be supported, but it should not be seen as more than an evolutionary step. The London Suppliers Group serves a useful purpose, but it lacks wide acceptance. For the time being we also need the special safeguards requirements of bilateral agreements, but through the creation of uncertain supply conditions these could prove counter-productive, encouraging some countries to attempt an otherwise unnecessary degree of nuclear self-sufficiency.

Thus, for some time now, there has been a wide acceptance of the need for other non-proliferation measures as well. They are referred to here as supplemental measures for convenience, although it does not follow that they should simply be added to existing measures without some mutual adjustment.

*The opinions expressed in this paper are those of the author and do not necessarily reflect the official policy of his professional affiliation.

II. A programme of supplemental measures

The provision of a programme of supplemental measures that are practical—and as adequate as can reasonably be expected—must be diligently sought after. The keynote should be a system whereby the country that claims to be using nuclear energy for peaceful purposes only can provide full and continuing assurance of this fact. The supplemental measures will remove, or mitigate to the maximum, the fear of others, thus reducing international tension and minimizing the causes for the development of nuclear weapons. Nuclear trade should as a result be freed from what is probably its greatest inhibition.

Such a programme might include the following: (*a*) a scheme for the international control of excess plutonium; (*b*) a scheme for the international control of excess highly enriched uranium; (*c*) a scheme for the international control of spent fuel; (*d*) joint participation by several countries, in sensitive processes, according to internationally approved standards which will enhance confidence in peaceful uses, and (*e*) finalization of existing proposals for physical protection.

Such a programme can be supported as follows: (*a*) it seems reasonably practical; (*b*) it would also be direct, offer a high degree of visibility in operation and not be dependent only upon promises and declarations, thus enhancing public acceptance, and (*c*) three of the schemes, those dealing with plutonium and spent fuel control, and that dealing with physical protection, are, at present, being examined under the auspices of the International Atomic Energy Agency (IAEA). The last-mentioned scheme is near to completion (IAEA, INFCIRC/225/Rev. 1). The scheme for control of highly enriched uranium probably would adapt readily from what is agreed in relation to the others. The fact that highly enriched uranium is not part of the commonly occurring fuel cycles would be an important consideration. Final results in relation to one or two of the schemes will not be achieved without detailed consideration and possibly difficult negotiation, but taken together they offer a reasonably sound solution and probably our best hope for the medium-term future.

Participation in the schemes relating to the control of materials should not be dependent upon membership of the NPT, but the need for IAEA safeguards, at least those relevant to the operations in question, should be retained. There are very few countries which have any objection to safeguards in relation to particular plants, materials or operations, and it is to be hoped that, in an appropriate nuclear environment, most of these countries would be prepared to modify their position. Control in each case could be operated by or under the IAEA, and schemes for multinational participation of the nature mentioned in (*d*) above could operate in accordance with the declared approval (in principle) of that body.

It is not necessary for every part of the programme to be introduced simultaneously nor for all participants to adhere to all the schemes. For the time being, some can have very limited application; not more than 12 countries have, at present, any form of reprocessing plant, and there are even fewer enrichment plants. It is desirable that the programme be in place as soon as is reasonably possible. However, it is better to proceed by a series of practical steps than to attempt a simple, short, comprehensive solution. If sufficiently well thought out, the schemes, taken together, should provide us with virtually all the confidence and protection we want. They can allow progressive development to meet any doubts or problems found to arise in practice. As to what we might finally arrive at, we should allow experience to be the guide and success the advocate.

Sanctity of agreements

Much has been said in the last year or two about assurances of supply. The matter was very fully examined in Working Group 3 of INFCE. It is likely that one or more of its proposals will be followed up. Most of the present concern about supply relates to the possibility of arbitrary, and what is seen as unjustified, termination of supply under current agreements. The concern is commonly considered as arising from the course taken by at least two countries in recent times. The basis of trade, international as well as domestic, is the agreement, and it is plainly necessary that agreements be carried out. *Pacta sunt servanda.* And this is not the less so when, from the purchaser's point of view, vital energy requirements are in question.

There must be some excuses for non-performance in international trade just as there are in intra-national trade, but recent experience has emphasized the desirability of these being strictly limited and of a reasonableness which commands the respect of both (or all) parties to the agreement. With long-term agreements it is rarely possible to anticipate all material events, or the effects of changed circumstances, but if there are special circumstances envisaged as a possibility at the outset by one party, which may lead it to abrogate the agreement or withhold supply or acceptance, it may well be desirable for the supply agreement to contain an appropriate clause. Intervention to abrogate or suspend such agreements is most likely to derive from governments, whose judgements are not always sound or objectively arrived at. There may be, and usually is, an agreement between the respective governments of the supplier and purchaser concerning nuclear supplies, and in that agreement it may be possible to say something appropriate to this problem. In any event, by whatever means (and there are many which suggest themselves), the possibility of arbitrary action should be minimized. Adherence to the programme outlined above could provide guiding criteria and would probably greatly reduce the likelihood of any such action. The need to have safeguards provisions of a duration beyond the reasonably foreseeable future would be obviated (see IAEA Gov. 1621).

III. Other considerations relating to nuclear trade

Nuclear wastes

Nuclear wastes, their disposition and the radioactivity problems associated with them should properly be subjects of concern to both suppliers and consumers. Australia, for example, produces some uranium now and will soon produce much more, but because of other resources, does not, for the time being at least, plan to produce nuclear power. It is quite opposed to the reception of nuclear wastes or spent fuel from other countries. Many of its citizens have been, and still are, of the view that Australia should not continue to mine uranium unless and until it is clearly shown that the waste problem has been solved. The government, accepting the finding of the Ranger Uranium Environmental Inquiry, does not take the view that the wastes position is at present such that it should ban the mining and export of uranium. It is obvious, however, that the treatment and disposal of wastes is an environmental consideration of global concern. It is also one which can have an important impact on trade in nuclear materials. In the long run, the management of wastes (and spent fuel) might become the critical consideration affecting the development of the industry.

Plant safety

Suppliers of a plant, such as a reactor, which can do serious damage if it proves defective or is incompetently managed, have a responsibility to the purchaser, and probably to other countries as well, to take all reasonable steps to maximize protection against any such result. What is reasonable will depend upon all the circumstances, including the training, competence and experience of the people into whose hands the reactor is delivered. Certainly, in the present climate of public opinion, a serious safety failure in one country can be expected to produce powerful anti-nuclear reactions in others.

Provision of technology

Associated with this consideration is what is called "the provision of technology". Countries which are in need of technical assistance in nuclear matters should, within reason, be given the help they need. Financial considerations are of particular importance, having in mind the huge sums usually involved in the purchase of plants. Third World countries commonly wish for, and should receive, technical assistance in a number and variety of areas; however, having in mind the public debate concerning nuclear development, what needs emphasis again is that nuclear technology is complex and cannot be delivered in lumps, like so many bags of rice. To be of any assistance, there needs to be in the Third World country (as with any other) a strong and sound infrastructure in all related areas. If the technology is not to be wasted, or accidentally misused, a large number of people must be thoroughly trained in many skills and highly specialized operations. It may take a country many years before it has developed a technical base from which it can contemplate taking a step towards the installation of nuclear power without, at least, being and remaining heavily dependent upon outside assistance.

There is a further aspect that applies to all planning, namely, the time factor. There are, of course, lead times, which are apt to be much longer than many expect, but the particular matter emphasized here is that discussions of need have to be related to the time at which, on a realistic estimate, the need is likely to arise. For example, it is a fact that at present and for many years to come the principal purchasers of nuclear equipment and material, including uranium, will be the nations which currently have a high degree of industrial development. Third World countries, with relatively few exceptions, will not need, or be able to use, nuclear plants or materials for power production for 10 or 20 years or more, and even then their requirements are likely to be small. They may in the meantime wish to make a start in the research field, but the reality of their need to do so will, among other factors, depend upon the likelihood of their future dependence on nuclear energy and when that dependence is likely to arise. I would hope that well before that time international arrangements, possibly along the lines of those already mentioned, will be in operation and will facilitate the necessary international transactions.

Paper 10. The role of institutional measures[1] in strengthening the non-proliferation régime

S. LODGAARD*

International Peace Research Institute, Oslo, Rådhusgata, 4, Oslo 1, Norway

I. The slowdown of nuclear power programmes

During the first half of the 1970s, the international nuclear market expanded rapidly in terms of the volume and value of transactions and the numbers of supplies and buyers involved. The first fuel cycle agreements were negotiated during that time. In the second half of the decade, however, commerce in nuclear materials, equipment and technology has been slowing down significantly.[2]

For various reasons, nuclear power plans have been considerably cut back. The estimate of the International Nuclear Fuel Cycle Evaluation (INFCE) for nuclear generating capacity in the world outside the centrally planned area by the year 2000 is in the range of 850–1 200 GW(e) (net),which is less than half of what was forecast only a few years ago. The number of Third World countries with firm nuclear power programmes —once expected to grow rapidly during this decade—has remained rather stable: 12 countries are now scheduled to have power reactors in operation by 1984. There have been only two newcomers over the past 10–15 years—Cuba and Iran, and the Iranian programme has now been brought to a halt. Certainly more countries are indicating an interest in nuclear power, but the pace is much slower than previously assumed.

The slowdown in the construction and spread of fuel cycle facilities is

---

*The opinions expressed in this paper are those of the author and do not necessarily reflect the official policy of his professional affiliation.

[1] The term 'institutional measures' is used in a broad sense consistent with the meaning assigned to it by INFCE. It is taken to include a wide range of undertakings by either governments or private entities to facilitate the efficient and secure functioning of nuclear power programmes and to reduce proliferation risks. It encompasses commercial contracts, technical assistance programmes, international studies, bilateral governmental agreements concerning supplies and non-proliferation assurances, as well as multinational and international institutions.

[2] See SIPRI (1979*a*).

of particular significance from a proliferation point of view. The technical difficulties at Tokai Mura and the plant modifications resulting from US–Japanese studies and negotiations mean a noticeable set-back for Japan's reprocessing plans. The same holds to an even greater extent for the Federal Republic of Germany, after the decision by the state government of Lower Saxony to postpone indefinitely the licensing of reprocessing at Gorleben. Other slowdowns in breeder research and development make it probable that only France and the Soviet Union will have substantial commercial breeder programmes by the turn of the century.

The export of fuel cycle components has also come to a temporary if not permanent halt. Some agreements have been cancelled, and the only remaining contract in this field is for the pilot reprocessing plant and the enrichment facility to be built in Brazil as part of the West German–Brazilian agreement. Much of the justification for these fuel cycle facilities depends, however, upon the magnitude of Brazil's nuclear power plans. Should its nuclear ambitions be reduced, the construction of the fuel cycle facilities may be affected as well.

By reducing the overall magnitude of the problem, the slowdowns may facilitate negotiation of restraints and make it easier for some of the countries involved to accept in practice restrictions that they are not ready to accept in principle (Dunn, 1979).

II. Unilateral export embargoes

On 16 December 1976 the French Council for External Nuclear Policy issued a declaration to the effect that, for the time being, sales of reprocessing technology would not be allowed. Following the French–West German summit meeting in Bonn on 16–17 June 1977, the government of FR Germany made virtually the same statement. The US Nuclear Non-Proliferation Act furthermore makes cut-off of US nuclear exports mandatory if an agreement for the transfer of reprocessing technology is entered into and discourages international commerce in enrichment facilities by threat of such cut-off, except in connection with INFCE or pursuant to an international agreement or undertaking to which the USA subscribes. The wording of the Act is such that it may well apply, for example, to West German assistance in the development of South African enrichment capacity, should this assistance continue and should there be political will to take action against it. The Soviet Union has no declared policy, but its record is uniform in denying exports of fuel cycle facilities after the breach with China in 1959.

In one respect, the French policy represents a variation: France will refrain from pressing other nations to forgo fuel cycle facilities which they may acquire indigenously or in co-operation with other suppliers (Lellouche, 1979). This is reasonably congruent with the *mondialiste* framework of French foreign policy: trying to reduce confrontation with the developing countries as a pioneer in the North–South dialogue. In the long run, supplier policies may again diverge, depending on the choice of fuel cycles and subsequent interest in international nuclear commerce. While Soviet policy seems firm and US policy is enshrined in law, West European suppliers are more likely to lift embargoes again, should efforts at reaching consensus fail.

So far, the coincidence of renewed interest in export restrictions and slowdowns in nuclear power programmes has had some significant non-proliferation benefits. Not only have these efforts reduced the spread of the capability to make nuclear weapons, but they have also provided some time for the rethinking of non-proliferation policies and have limited the magnitude of the problem which the interconnections between civilian and military applications of nuclear energy create.

III. The safeguards situation

The Zangger Committee 'trigger list' was finalized in August 1974, stimulated by the Indian test of 18 May that year. The trigger list of the London Nuclear Suppliers Group added a few more items in 1976, and the Zangger list was recently updated so as to become virtually identical with that of the London Group. Pakistan's recent circumvention of nuclear export controls has raised the question of further extension of the list. As a means of reducing the grey market for nuclear equipment and technology, key grey-area components might be identified and added. Supplier nations might furthermore review their export regulations with the aim of tightening the controls on ambiguous items. Improved intelligence monitoring, both nationally and through international co-operation in intelligence gathering, would serve the same end.

Today, the effectiveness of national export regulations varies considerably. Much can be done to tighten the controls, and limitations of the grey market ought to be a fairly consensual matter among supplier states. Due attention should furthermore be directed to the relationship between the scope and nature of the grey market on the one hand, and the rules of the open market on the other: tightening of export controls may be constructively combined with some relaxation of restraint on open transactions.

The demand for full-scope safeguards as a condition for transfers of nuclear materials, equipment and technology to other countries—both parties and non-parties to the NPT—is a basic non-proliferation requirement.

The East European countries have declared their willingness to adopt this policy as soon as all members of the London Group are ready to do the same. Later, however, the credibility of their position has been somewhat shattered by Prime Minister Kosygin's promise to deliver enriched uranium for India's Tarapur power station if and when US deliveries are brought to a halt because of Indian refusal to bow to the US full-scope safeguards demand (*Nuclear Engineering,* 1979).[3] It remains to be seen whether US policy will actually be undercut in the name of regional power politics. So far, however, unanimity in the London Group has hinged on France and FR Germany. The pressure on these countries to abide by the full-scope safeguards principle is mounting before the next NPT Review Conference, and supplier agreement may now be within reach.

Outside the five nuclear weapon states recognized by the NPT, 12 operating nuclear facilities in five countries (Egypt, India, Israel, South Africa and Spain) are presently not subject to IAEA (International Atomic Energy Agency) or bilateral safeguards. In addition, there are some unsafeguarded laboratory-scale activities such as the reprocessing facilities in Egypt and Pakistan and a small-scale fuel fabrication capability in Pakistan. Pakistan's latest advances towards a centrifuge enrichment facility are certainly outside the safeguards domain as well.

The unsafeguarded plant in Spain—a power reactor operated jointly with France—is not very significant. Spain will join the NPT, or at least accept full-scope safeguards when it enters the European Communities. The Inshas research reactor in Egypt, obtained from the Soviet Union around 1960, is too small to be of real significance. The three remaining countries with unsafeguarded facilities—India, Israel and South Africa—are the real problems. The power programmes of India and South Africa depend on outside support—primarily deliveries of enrichment services from the United States—so the supplier states therefore have some leverage in threats of cut-off. The pressure is on: the full-scope safeguards demands of the US Nuclear Non-Proliferation Act took effect on 10 September 1979.[4] The Israeli programme is more exclusively military in nature and is largely self-sustained.

Supplier agreement on full-scope safeguards is essential for regional arms control. A regional arrangement to restrain the nuclear competition in South Asia can, for instance, hardly be conceived of without international safeguards on all nuclear activities in the countries involved. In the case of South Africa, imposition of full-scope safeguards may not only be instrumental in undermining its nuclear weapon ambitions, but may be the

[3]The promise was made during Mr Kosygin's visit to India in March 1979. However, the joint communiqué of 15 March merely refers to co-ordination between the two countries in the field of peaceful uses of nuclear energy without going into specifics, and Soviet sources have later indicated some modification of the Prime Minister's statement, hinting that Soviet policy may actually be more cautious and restrictive.

[4]Pending applications may still be met until March 1980 (two years after the Act entered into force), while new applications will be turned down unless full-scope safeguards are accepted.

only way to prevent South African sales of highly enriched uranium in the medium-term future, offered for a suitably high economic, political or military price.[5] South Africa may well have bowed to the demand when the Koeberg contract was concluded in 1976, had the London Group agreed on it. Today, it seems somewhat less likely. India is well known to be staunchly anti-NPT and is challenged both by China and Pakistan; even if the suppliers refrain from undercutting each other, India's acceptance will therefore be very hard to obtain.

However, *de facto* application of safeguards to all facilities on the territory of a non-nuclear weapon state is not enough. The full-scope safeguards requirement also implies the legal obligation to declare construction activities to the IAEA for design review and subsequent application for safeguards. Non-NPT countries other than those mentioned above may therefore turn down the full-scope safeguards requirement as well. The best prospect for acceptance of full-scope safeguards by the Latin American countries in this category—Argentina, Brazil and Chile—lies in unqualified adherence to the Treaty of Tlatelolco. So far, Argentina has not ratified the Treaty, and it is not yet in force for Brazil or Chile. Some moves have recently been made, however, towards full application of its provisions. Should it be fully accepted, Latin America would be the first nuclear weapons-free zone in an inhabited part of the world.

IV. The supply problem and autarkic fuel cycle programmes

The US Nuclear Non-Proliferation Act also defines a number of other conditions for the export of nuclear items, some being similar to those agreed upon by the London Group and others different. Among the most important conditions are the requirement for prior US approval of reprocessing, enrichment, other alteration and storage of materials of US origin. The London guidelines merely recognize the importance of mutual agreement on such matters whenever appropriate and practicable. Australia and Canada require much the same safeguards and prior approval clauses as does the USA.

In the absence of supplier agreement, commercial competition between countries insisting on full-scope safeguards and countries that require application of safeguards to the exported items only continues, to the detriment of non-proliferation. As a rule, the former are also more demanding about other export conditions. Importing countries may therefore turn to European suppliers for equipment and technology: to South Africa and France/Gabon/Niger for uranium, and to

[5] See Lodgaard (1979).

Eurodif/Coredif, Urenco and possibly the USSR for enrichment services. Accordingly, European countries may prefer Australian, Canadian and US uranium for their domestic power programmes, so that enough South African uranium is available to support reactor orders from countries which do not fulfil the strictest requirements.[6] For countries not accepting full-scope safeguards, this is the road for the time being. Conversely, West European countries are close to having a monopoly on sales of equipment, technology and services to these countries, and thus have a strong commercial interest in maintaining the present situation.

For Third World importers, therefore, the supply problem is not only a question of the adequacy of world production capacities, but to a large extent also a question of the availability of supplies. The reduction of nuclear power plans has to some extent alleviated the first type of concern. INFCE found that the uranium industry should not experience undue difficulties in meeting requirements up to the year 2000, provided that the necessary exploration and investment can be made. It also concluded that present enrichment capacities under operation or construction would cover projected enrichment needs until around the year 1990, whereas adding presently planned capacities would cover projected needs until somewhere between the years 1995 and 2000.

The availability of supplies as determined by export policies has, however, become more of a problem. Being confronted with a combination of embargoes and transfer conditions which in some cases involve foreign governments in the day-to-day operating decisions of their nuclear power organizations, non-aligned countries may not only use the special import road outlined above (as long as it exists), but may also try to increase their nuclear self-sufficiency, following the examples of India and Argentina. They may increasingly seek national solutions to their fuel cycle problems and turn to each other for mutual assistance. The proliferation implications of higher levels of nuclear independence are potentially grave.

In trying to halt or restrain developments in this direction, the complementary nature of supply assurances and non-proliferation guarantees must be fully recognized. INFCE emphasized that not only do effective non-proliferation assurances facilitate stable and predictable supplies, but a country's non-proliferation commitment may also be considered stronger the greater its reliance is on the international market for nuclear supplies. In particular, better supply assurances may reduce the incentives for acquisition and spread of reprocessing and enrichment facilities. The more unstable and the less predictable the international market is, the stronger the pressures for autarkic programmes are likely to be. Small-scale fuel cycle activities may be promoted as a mobilization base in case supplies fail—or the rationale can be used as a pretext for programmes of military intent.

[6]To what extent Western Europe is organizationally equipped and able to execute such a policy is another question.

While the full-scope safeguards requirement is a welcome feature in the nuclear policies of supplier states, and universal adoption of this principle must be continuously urged, the unilaterally imposed export conditions that apply at present may have some unfortunate implications in the long run. It is therefore vitally important that INFCE is followed by joint consumer–supplier talks on mutually acceptable restraint.

V. Extending the non-proliferation régime

Consensus negotiations will have a built-in tendency to extend the coverage of the non-proliferation régime at the expense of the proliferation resistance of its provisions. Modification of restrictions may be traded for more universal adherence.

Over the long term, there is a strong technical argument in favour of such a trade-off. The most inevitable dissemination of existing and new technologies will in any case make access to fissionable materials easier: if the already widespread research on, for example, laser enrichment comes to fruition, it could greatly facilitate the production of highly enriched uranium.[7] In the long run, therefore, extension of the coverage of the régime might be given priority over the uphill fight for a fire-break between civilian and military applications.

INFCE can be taken to support this conclusion. It considered that technical remedies to proliferation were a question for the next 20–25 years—not because the proliferation problem would disappear, but because the spread of expertise and equipment gradually renders technological fixes meaningless. And INFCE was not very successful in devising significant technological barriers that could have large-scale application before the turn of the century.

Other factors add urgency to extending the coverage of the régime. Third-tier exporters which are partly outside the present régime and therefore free of some of its key obligations are about to appear on the international market (Dunn, 1979).[8] India has been training Egyptian, Iranian and Vietnamese nuclear scientists and may be capable of exporting not only large research reactors but also small power reactors; Argentina has sold a research reactor to Peru and renders nuclear assistance to Chile, Colombia, Ecuador and Paraguay; Spain is assisting Ecuador in building a research reactor and has entered nuclear assistance agreements with Colombia, Israel

[7] See SIPRI (1979*b*).

[8] First-tier suppliers comprise established nuclear weapon states such as France, the Soviet Union and the United States, while second-tier ones include, for example, Canada, FR Germany and Sweden.

and Venezuela; South Africa may soon appear on the international market as a supplier of enrichment services; and during the 1980s, Brazil, Israel, South Korea and Taiwan may also come into a position to transfer nuclear equipment and technology to other states. Most of the third-tier exporters are not bound by any trigger list (all except South Korea and Taiwan which, on the other hand, may not be considered the firmest members of the NPT). The London guidelines strengthened the control of retransfers, but they do not cover sales or transfers that are not based on technology acquired from members of the Group.[9] Gradually, the emergence of such third-tier exporters will undermine the non-proliferation régime, if they cannot be made to join it.

In trying to extend the coverage of the régime, prior-consent provisions might have to yield. Bilateral agreements requiring prior consent on reprocessing might be relinquished in favour of a system of international control. An international entity or arrangement might substitute for bilateral reprocessing controls, and the bilaterals could be reduced to fall-back options. This might, in a sense, be analogous to the assignment of bilateral inspection rights to the IAEA in the beginning of the 1960s, and the fall-back option in the bilateral agreements would resemble the strengthening of bilateral safeguards (as a last resort option) in some of the recent co-operation agreements.[10]

The most restrictive prior-consent policies might have to yield anyhow. A number of countries are moving ahead towards plutonium recycle, particularly in breeders, not necessarily because this is the optimal economic choice, but because large investments in the uranium–plutonium cycle are not easily abandoned in favour of much less-developed fuel cycle concepts. INFCE seems, moreover, not to have identified major, new alternatives for the short- or medium-term future and furthermore concludes that no fuel cycle should be discounted on proliferation grounds alone if there are prudent economic and energy-policy arguments for implementing it. Accordingly, the bulk of the work at INFCE was devoted to the identification of technical or institutional measures to reduce the proliferation risks of various fuel cycles, without indicating any general preference for one over the other.

How far the United States is willing to go in accepting reprocessing and use of plutonium in breeder reactors is unclear. There seems, however, to be a growing recognition among US policy-makers that for some countries, plutonium breeders would have to be accepted. On the other hand, for a variety of reasons—the slowdown in nuclear power plans, technical difficulties with reprocessing plants and breeders, associated environmental dangers and proliferation concerns—the advent of the plutonium economy

[9] Paragraph 10 of the London guidelines. For an evaluation of the guidelines, see SIPRI (1977).

[10] The strengthening of bilateral safeguards as a fall-back option appears, for example, in recent Canadian nuclear co-operation agreements. See also Gilinsky (1978).

will be much slower than anticipated a few years ago. And largely due to US emphasis on the proliferation risks of plutonium utilization, a number of measures to reduce those dangers are now being seriously considered.

An international arrangement for plutonium control

The scheme for International Plutonium Storage (IPS), now being elaborated at the IAEA, might form the nucleus of an international arrangement to substitute for the bilateral prior-consent provisions. According to Article XII.A.5 of its Statute, the IAEA has the authority to operate such a scheme, and the creation of an IPS scheme has been encouraged by INFCE. Establishment of an IPS scheme would imply that excess plutonium has to be deposited with internationally controlled stores at reprocessing and possibly fuel fabrication plants, and can be released for specified peaceful purposes under international safeguards only. The criteria for release would be non-discriminatory and concrete, so as to avoid ambiguity and uncertainty. Most release decisions might therefore be effectively resolved, ensuring prompt delivery. Retransfers of released plutonium would not take place without the consent of the IPS controlling machinery.[11]

The IAEA Expert Group on International Plutonium Storage has agreed that the stores may be located both in nuclear weapon and non-nuclear weapon states, and that the scheme should be acceptable to as many countries as possible. However, if the scheme is to be negotiated within a wider framework—as an integrated part of a more comprehensive, reinforced supply assurance/non-proliferation régime—only NPT members, or countries accepting similar obligations, may be considered eligible for membership (possibly, acceptance of full-scope safeguards could suffice).[12]

Viewed in isolation, the IPS scheme has no great appeal to those who have not yet entered into the fold. Rather, it is designed to keep temptation away from the virtuous (Fischer, 1979). But if it is combined with abolishment of present prior-consent demands, participation in the scheme could have the attraction of securing uninterrupted access to plutonium fuels. Thus combined, the arrangement would stand out in clear contrast to the present US policy of granting MB-10s on a case-by-case basis (primarily for lack of reactor storage capacity). This case-by-case approach is seen by some as a blatant formula for uncertainty, whereas the IPS criteria for release could be clear to the extent of leaving the bulk of the decisions to the IPS administration.

---

[11] See reports from the IAEA Expert Group on International Plutonium Storage.

[12] On British initiative, a full-scope safeguards model agreement has been developed at the IAEA which provides for safeguards as adequate as the NPT arrangement, but which do not presuppose acceptance of the Treaty. So far, however, the model agreement has not been applied anywhere.

The non-proliferation effectiveness of the arrangement may be enhanced by co-location and co-conversion. Location of different fuel cycle facilities, for instance, reprocessing and MOX-fabrication plants, on the same site is recommended by INFCE as one of the most attractive non-proliferation measures in the short-term. The same largely goes for co-conversion—the production of mixed oxide from uranium and plutonium solutions—where a modest amount of further development effort is needed before industrial-scale application can take place.

A prohibition of the release of pure plutonium would further enhance the effectiveness of the scheme. It has therefore been suggested that plutonium should only be released from international storage in the form of mixed oxides of plutonium and uranium. Agreement on this would logically apply also to the transport of plutonium from a reprocessing plant to a fuel fabrication plant at a different site. Since co-processing is a more distant and uncertain prospect, this problem can best be solved by a policy of co-location.

INFCE has also encouraged further study of an international arrangement for storage and management of spent fuel. An international spent fuel arrangement might have some non-proliferation relevance in helping to come to grips with the 'plutonium mine' problem. It could furthermore make the once-through cycle more attractive. The release of spent fuel for reprocessing is, however, likely to be based on much the same criteria as those that determine the release of plutonium from an IPS facility. It therefore yields no great non-proliferation gain beyond that represented by the IPS. Basically, international management and storage of spent fuel are a response to an urgent technical problem rather than to a non-proliferation concept.

Restricting the number of reprocessing plants

One international plutonium store at each reprocessing facility would at present amount to some 10 sites, or about a dozen under ideal circumstances of Indian and Argentinian participation. The number should be kept as low as possible.

Multinationalization of reprocessing operations would serve that end. It offers economies of scale, supply assurances and non-proliferation benefits. The non-proliferation benefit would be a product of the reduced spread of plants and the multinational control of their operation.

There is a wide range of possible forms of multinational involvement in fuel cycle plants. Today,, advance payments for reprocessing services at La Hague and Windscale cover part of the construction costs of these plants. Such debt financing could be made in return for a certain share of the reprocessing capacity or some guaranteed long-term services. Equity sharing, as practised by France in the enrichment sector, may be a more appropriate form of involvement to this end. Equity investment could lead to multinational participation in the management of the plant as well and therefore enhance the effectiveness of safeguards. At INFCE, a step-by-step progression towards multinational policy determination and joint operation of fuel cycle facilities was envisaged. In this way, concentration of reprocessing capabilities at some few, large plants could be encouraged, and the further spread of reprocessing plants could be reduced.

However, some radical proposals should be kept alive as well. In many ways, the best arrangement for supply guarantees and non-proliferation assurances would be the management of all sensitive fuel cycle facilities under the authority of an international agency, starting with the submission of existing plants to the agency. Such a proposal may not seem very real, however, and indeed may never be effected. Scientists as well as politicians ought to know, however, that the merits of a proposal are not limited to the prospects of realizing it. Proposals usually have other important functions as well. To ask for 'much' is sometimes a sensible strategy for achieving 'something'.

Another way of limiting the spread of plutonium would be for Western Europe, the Soviet Union and Japan to avoid the temptation to reduce unit costs for plutonium breeders by premature exports and to limit commercialization to situations where there are compelling advantages. On the importing side, the demand for breeder technology may be kept low, well into the 1990s, if a high degree of openness about breeder research and development—or access to breeder technology—could be guaranteed. France, the Soviet Union, the UK, and eventually Japan will ensure a high degree of supplier diversification and commercial competition should plutonium breeders prove clearly advantageous for other countries at a later stage. In that case, the United States might also become a supplier of breeder equipment and technology.

Policies to limit the application and spread of breeders should concentrate on positive sanctions, that is, incentives and encouragements, rather than on (supplier) efforts to institute prohibitions. Quite a few countries still consider the plutonium breeder a key option for guaranteed energy supplies and greater energy independence in the long run; and in such an environment, negative sanctions may do more harm than good.

Supply assurances

Plutonium recycle in breeders or thermal reactors can also be delayed and avoided by better and more predictable access to uranium fuels for use in the once-through mode and by more effective means to handle reactor waste. In executing its non-proliferation policy, the Carter Administration has so far failed to bring forth the specifics of the promised incentive programme for fuel assurance and waste management.

There is broad agreement that, apart from problems caused by government intervention, the commercial market has so far been reasonably successful in providing nuclear fuel, equipment and technology. INFCE reiterated the belief that the commercial market will continue to be the major instrument for ensuring supplies to nuclear power programmes.

The market should allow for consumer protection by diversification. This is largely the case today: the number of suppliers is significant, as well as their political, economic and other divergencies. Consumer protection can be further enhanced by offering a variety of contractual terms and conditions, by establishing a sound market for spot transactions, and by freer exchange of information. Importers should not have to place all their eggs in one basket, and the market already offers good opportunities for distribution of risks.

Long-term supplies of uranium, enrichment and fuel-fabrication services can, furthermore, be ensured by consumer contributions to the equity or debt financing of the supplier enterprises—as mentioned above for the reprocessing sector. INFCE also gave consideration to the establishment of a 'uranium emergency safety network' among utilities, built on existing practices of fuel loans in cases of market failure. The pursuit of an international nuclear fuel bank—a statutory obligation for the President of the United States—might furthermore be of some particular interest for nuclear power newcomers and countries with small nuclear programmes.

The NPT makes it a government obligation to facilitate the fullest possible exchange for the peaceful uses of nuclear energy. While the commercial market is likely to remain the major supply instrument for the foreseeable future, governmental back-up systems may therefore not only be appropriate, but can become a legitimate demand by consumer states. The most urgent task in this field is, however, to clarify and facilitate the access to nuclear items by elaboration of mutually acceptable conditions for supply between importers and suppliers. Uniform and predictable application of jointly elaborated criteria would mitigate uncertainties, and prescribed procedures for joint management of changes in non-proliferation clauses might further reduce the fears of supplier interference.

Technology denial and safeguards

The supplier policies of technology denial were not among the major subjects of INFCE. It has usually been assumed that the French and West German export bans would be in force till INFCE came to an end, but that beyond INFCE, they may be less firm. New technologies—some more and others less proliferation-resistant than those in use today—will, moreover, enter the commercial stage and therefore raise new questions of export policy as well. As noted above, embargoes are furthermore likely to have some counter-productive effects in the long run. It is therefore essential that they be reviewed and reconsidered in the near future.

The review of export restraints and embargoes should take place in the wider context of reconsiderations regarding supply assurances and non-proliferation guarantees. It belongs to the broader issue of the operative contents and implications of Articles III and IV of the NPT. Four different positions might be considered: (*a*) to maintain the present embargoes backed up by sanctions (the US policy); (*b*) to maintain the embargoes but without sanctions (the French policy): (*c*) to allow sales of sensitive fuel cycle technologies to multinational or international projects only, or (*d*) in principle, to be willing to sell a wider spectrum of technologies to all countries accepting full-scope safeguards, the IPS and other elements of a strengthened non-proliferation régime. If consensus negotiations on mutually acceptable mechanisms for supply assurance and proliferation restraint begin, sales of sensitive fuel cycle technologies should probably be allowed only to multinational and international enterprises during the course of the negotiations.

In broad terms, a softening of embargoes could be thought of in combination with better export controls to reduce the grey market, longer trigger lists, improved safeguards and stepped-up sanctions against countries violating non-proliferation obligations.

A number of safeguards issues constitute integral parts of the supply-assurance/proliferation-restraint discussion, in any case. INFCE considered some new concepts and measures for safeguarding of reprocessing and MOX-fuel-fabrication plants, based on real-time material accountancy and augmented by new techniques of containment and surveillance as well as design criteria for effective application of safeguards. INFCE did not respond, however, to the growing need to clarify what the basic function of safeguards should be. So far, two different notions seem to exist side by side. The one holds that the function of safeguards is to verify the observance of international commitments concerning non-proliferation of nuclear weapons. The other adds the function of timely warning—publicity before consummation—to it: timely detection of significant quantities of fissionable materials should give the international community sufficient warning time to take action before the explosive device is an established fact. In the future, some further reduction of lead time between civilian and military applications can hardly be avoided. It therefore seems desirable that the function of safeguards be explicitly discussed in the aftermath of INFCE.

VI. Conclusions

INFCE reaffirmed the importance of the NPT, of international safeguards and of the role of the IAEA. An evolutionary process towards a more transparent, comprehensive and integrated supply-assurance/non-proliferation régime was envisaged. An evolutionary process based on existing instruments and institutions is, however, not necessarily tantamount to *incrementalism*. In much the same way as the NPT contains a bargain involving the safeguards obligations of Article III and the supply obligations of Article IV, further improvements of the non-proliferation régime may have to be conceived of in terms of packages. A step-by-step evolution of the régime would, anyhow, have small chance of success if the single step is not designed with due consideration to the overall distribution of rights and obligations.

INFCE was set up as a technical study rather than a negotiation. It nevertheless prevented the main parties concerned with the plutonium issues from drifting further apart. The North–South controversy over export policies and safeguards regulations would probably have developed further without INFCE. Thus, a certain momentum was created that ought to be upheld and reinforced by joint supplier–consumer talks geared towards mutually acceptable mechanisms for supplies and proliferation restraint. It is important that those affected by the rules participate in the generation of them: this way, it becomes more likely that the rules will have the intended effects. Consensus negotiations are difficult, but may yield better results in the long run. The rules and guidelines thus elaborated might evolve into a separate international treaty or perhaps an additional protocol to the NPT defining the operative meaning and implications of Articles III and IV.[13]

In addition to the subjects covered by INFCE, the talks should comprise the issues of technology denial and the functions of safeguards. The IAEA has proved that it can provide adequate machinery for consideration of many of the issues involved. Its Board of Governors may discuss where to go from INFCE—ways and means of carrying considerations further on the political level—giving due attention to the second Review Conference of the NPT and to the prospects for another conference on the peaceful uses of nuclear energy, planned at the initiative of the non-aligned states.

However, while successful consensus negotiations may be necessary to bring more countries into the fold, it may suffice only in rare and less

[13] Article IV of the NPT must be seen in relation to Article I as well. For a discussion of this relationship see Wohlstetter (1978). The legislative history of the IAEA Statute shows that "peaceful" was intended to mean "exclusively peaceful"; similarly, it was implied by some that the term "peaceful uses" in Article IV of the NPT should be taken to mean "exclusively peaceful", and not military. Today, this distinction is widely recognized as untenable. In the meantime, however, "peaceful uses" has, in essence, been made a matter of intent: proscribed activities were those that could not be said to have a meaningful peaceful purpose or intent, a wide concept which went far beyond activities having a clear economic rationale. In a sense, there is nothing surprising or spectacular about this: nations sometimes want to improve their research and development capabilities without any fixed practical application in mind. Options which are considered promising for some reason or other are pursued to explore what they can deliver in the longer term, or for the prestige they may confer. Both civilian and military nuclear programmes are still considered status projects in many parts of the world. However, while persistent efforts are made to reiterate the original interpretation of Article IV in the face of present export restraints and embargoes, reconsideration of international market rules on the basis of proliferation-resistance criteria is now gaining wide acceptance as well. Largely due to INFCE there is, in a sense, broad support for redressing the balance between Articles I and IV of the NPT so as to keep the proliferation risks of international transfers as low as possible.

interesting cases. For significant régime extensions to take place, military–political measures are needed.

References

Dunn, L., 1979. Half past India's bang, *Foreign Policy* 36:71–89.

Fischer, D.A.V., 1979. Role of existing international arrangements and institutions, Lecture at the ANS Executive Conference on International Nuclear Commerce, 11 September.

Gilinsky, V., 1978. International discipline over the uses of nuclear energy, in Wohlstetter, A. *et al.* (eds), *Nuclear Policies: Fuel Without the Bomb* (Ballinger, Cambridge, Mass.).

Lellouche, P., 1979. France in the international nuclear energy controversy: a new policy under Giscard d'Estaing, *Orbis* 22 (no.4), Winter.

Lodgaard, S., 1979. *Nuclear Collaboration with South Africa: Cut-off, Full-scope Safeguards, or Extension of the Non-Proliferation Treaty?* PRIO Publication S-38 (PRIO, Oslo).

Nuclear Engineering International, 1979. 24 (no. 285), p.4.

SIPRI (Stockholm International Peace Research Institute), 1977. *World Armaments and Disarmament, SIPRI Yearbook 1977* (Almqvist & Wiksell, Stockholm), Appendix 1A.

SIPRI, 1979*a*. *World Armaments and Disarmament, SIPRI Yearbook 1979* (Taylor & Francis, London), pp. 305–28.

SIPRI, 1979*b*. *Nuclear Energy and Nuclear Weapon Proliferation* (Taylor & Francis, London), paper by K.L. Kompa, pp.73–90.

Wohlstetter, A., 1978. In Wohlstetter, A. *et al.* (eds), *Nuclear Policies: Fuel Without the Bomb* (Ballinger, Cambridge, Mass.), pp. 40–42.

Paper 11. Energy independence via nuclear power with minimized weapon-proliferation risks

K. HANNERZ*

Asea-Atom, Box 53, S-721 04 Västerås, Sweden

I. Introduction

Internationalization of fuel cycle facilities is undoubtedly one of the best proposals for realizing the nuclear power benefits to a great number of nations, while avoiding concomitant proliferation of the capability to produce nuclear weapons. Efforts towards achieving this goal, on a world-wide scale, should receive high priority. However, unless this work is based on a realistic appraisal of the motivation of most nations for using nuclear power, it risks being ineffective or even counter-productive. The strongest motivation for an extensive commitment to nuclear power is energy independence.

Countries without domestic uranium resources quite naturally have a limited commitment to nuclear power as a component of a suitably diversified mix of energy-supply sources. For these nations, nuclear power is not a path to energy independence. Rather, few objections can be raised to reliance on fuel cycle facilities outside of national control. Energy independence would have to be based on some source other than nuclear power, or it would be impossible altogether. The situation would be different if sizeable domestic uranium resources were available. In the latter case, taking the course towards maximum independence of outside supply would represent an almost irresistible temptation, particularly in countries with a strong nationalistic trend in their overall political stance. It is precisely this category of nations that represents the real problem in terms of nuclear weapon proliferation.

In pursuing a policy towards internationalization of fuel cycle facilities, the struggle towards independence must be accepted as both a fact of life and a perfectly respectable approach. Otherwise, opposition, rather than acceptance, will result among those nations who are important in this context.

---

*The opinions expressed in this paper are those of the author and do not necessarily reflect the official policy of his professional affiliation.

It is at this point that a sharp distinction should be made between the two main types of fuel cycle activities under consideration, namely enrichment and reprocessing.

II. Comparison of enrichment and reprocessing

For those nations using light water reactors, enrichment is, of course, an absolute requirement for fuel supply. Reprocessing, on the other hand, is definitely not a requirement for nations using thermal reactors, whether they are light or heavy water moderated. (We can disregard high-temperature reactors in this context.)

Reprocessing

In recent years, economic comparisons have been made in Sweden between a fuel cycle using reprocessing, plutonium recycle and final disposal of vitrified waste, and a cycle simply involving irradiation and subsequent final disposal of the fuel without reprocessing. The calculations were based on actual contract prices for reprocessing and on the most complete and detailed proposals that have thus far been worked out anywhere in the world for environmentally acceptable final disposal of both kinds of waste.

The conclusion was that the reprocessing route would be the more expensive even with today's high uranium prices. For this reason, claims that a domestic reprocessing plant is a necessity for achieving energy independence cannot be based on economic reality. Such claims must be regarded with distrust and as an indication that the real purpose behind them is probably a desire to procure, or at least to acquire the option to procure, nuclear weapon material. Indeed the indications are that world-wide reprocessing on a really major scale may be deferred until the advent, in earnest, of the breeder and that plutonium recycle in thermal reactors will take place only to a limited extent, if at all.

Clearly a strong approach to internationalization of any reprocessing plants needed beyond those already operating or committed to construction in France and in the UK is fully warranted.

Enrichment

The situation will be very different if enrichment technology remains permanently beyond national control. Nations seeking energy independence are then strongly discouraged from relying on light water reactors (LWRs).

Rather, they are expected to select heavy water reactors (HWRs) with refuelling during operation, which is the current trend and has been for some time.

There can be no doubt that such HWRs present a much greater problem than LWRs from the proliferation point of view. The reason is that HWRs can be used for the production of weapons-grade plutonium without significant interference with power generation. There are good possibilities for avoiding detection unless the international inspection programme is extremely effective. Also the weapons-grade material can be obtained in a reprocessing plant that is relatively simple, compared to one needed for reprocessing high burn-up LWR fuel.

III. Conclusions

Aside from the objections based on national commercial aspirations, everyone must agree that the widespread construction of reprocessing facilities in so-called nuclear threshold countries should be avoided and the present trend in that direction stopped.

On the other hand, the widespread construction of national enrichment plants (as an alternative to proliferation-prone heavy water reactors) is at least as undesirable if they can be converted to supply highly enriched, weapons-grade uranium (such as would be the case with a centrifuge plant). It is in this connection that the French proposal for a proliferation-resistant enrichment technology announced at Salzburg in 1977 offers so much promise. As has been made public, an agreement has now been reached between France and the USA for further development and evaluation of this method.

There seems to be little doubt that a world order with a number of such national proliferation-resistant enrichment plants and light water reactors supplied by them (with everything under IAEA inspection) is much to be preferred from the non-proliferation point of view to a system involving a large number of heavy water reactors with on-load refuelling (provided, of course, the French claims of proliferation resistance can be substantiated).

It is for this reason that the indiscriminate insistence on internationalization of all fuel cycle facilities is bound to be counter-productive in terms of preventing the proliferation of the production capability of nuclear weapon material. Such an approach should be replaced with one with strong emphasis on preventing national reprocessing plants and with active encouragement of the development of a proliferation-resistant enrichment technology.

Paper 12. A nuclear fuel cycle pool or bank?

M. OSREDKAR*

Jožef Stefan Institute, Edvard Kardelj University, Ljubljana Jamova 39, 61000 Ljubljana, Yugoslavia

I. Introduction

The term 'nuclear fuel bank' has regularly appeared in recent years during discussions related to the use of nuclear power and nuclear fuel. Whatever is meant by the term, the motivation behind such a proposal may not in all cases be a strengthening of the non-proliferation régime but rather the promotion of nuclear power; in the opinion of some, the main function of such an institution should be to provide nuclear fuel to those in need of it. This difference in motivation is considered by some to be quite significant and is a reflection of other differences in attitude towards international arrangements such as the Non-Proliferation Treaty (NPT). As indicated by Goldschmidt and Kratzer (1978), the two attitudes are reflected in the two declarations made 23 years apart by Presidents Eisenhower and Carter, respectively, which are at the base of the two schools of thought prevailing today.

In this paper an account will be given of four different approaches to a 'nuclear fuel bank', reflected, in chronological order, in (*a*) the Statute of the International Atomic Energy Agency (IAEA), (*b*) the suggestion made by Yugoslavia at the Twentieth General Conference of the IAEA that the possibilities and ways of creating an international fuel cycle pool be studied; (*c*) the Nuclear Non-Proliferation Act of 1978 (NNPA), and (*d*) the International Nuclear Fuel Cycle Evaluation (INFCE) discussions. Finally, these four approaches will be commented on.

II. The Statute of the IAEA

The basic assumption behind the IAEA Statute is the international accep-

*The opinions expressed in this paper are those of the author and do not necessarily reflect the official policy of his professional affiliation.

tability of the promotion of the peaceful usage of nuclear energy under international control, as stated in Article II of the Statute. Once the nuclear weapon states realized they could not interfere with each other's interest in manufacturing nuclear weapons, they were able to agree on establishing the IAEA as it is now and to foresee a very wide range of uses for it. From the point of view of the present discussion, some of these are of particular interest. Among other things, the Agency already acts as an international broker in arranging for supplies of nuclear materials, equipment and facilities. Moreover, the Statute provides the institutional framework for the IAEA to operate as an international nuclear fuel bank (Articles III.A.2 and B.3, IX, XI, XII, XIII, XIV.B.2 and E to G), although the Agency has never played this role and the relevant provisions of the Statute have never been applied. Should there be a willingness to make use of it, the Agency already has the basic institutional mechanism and relevant expertise to be able to function as a nuclear fuel bank, once the member states or other interested groups have worked out the details. Whether this could, in fact, be achieved in view of the differences in interest and status of the member states in relation to nuclear power—differences that are strongly reflected in the Agency's organs—remains an open question.

III. The international fuel cycle pool

In September 1976, at the Twentieth General Conference of the IAEA held in Rio de Janeiro, Brazil, proposals were made by Yugoslavia along the lines of resolutions adopted at the fifth conference of the non-aligned countries in 1976. Yugoslavia proposed that the scope of the study on regional reprocessing centres be broadened to include not only the tail-end but also the front-end of the fuel cycle. It was suggested that reprocessing, which for non-proliferation reasons the USA was in favour of being done in regional centres, was at present for most developing countries a problem of less urgency than other parts of the fuel cycle. Yugoslavia further suggested that the possibilities and ways of creating an international fuel cycle pool be studied. The members of such a pool could make different contributions. Third World countries could contribute, for instance, natural resources such as ore or yellowcake. Developed nuclear countries, on the other hand, could contribute money or materials to the pool with the aim of accelerating the transfer of technology. Developing countries would thus be able to exchange materials or receive loans from the pool for investments and expenditures related to the fuel cycle.

The proposal (which was later reiterated) was noted, but a realistic evaluation of the possibilities of creating a fuel-pooling organization or a bank was never carried out, either by members of the Agency or by other groups. During the INFCE discussions, however, related ideas came up on several occasions; for this reason, it is mentioned here.

IV. The US Nuclear Non-Proliferation Act of 1978

The US Nuclear Non-Proliferation Act (NNPA), Section 104, authorizes the President to seek negotiations with a view towards the eventual establishment of an international fuel authority with responsibility for ensuring fuel supply on reasonable terms. Proposals would be made for initial fuel assurance including the creation of an interim stockpile of uranium enriched to less than 20 per cent.

The concept of an international nuclear fuel agency as described in the NNPA might easily provide the basis for some type of nuclear fuel bank, although the Act does not give any indication of how it could be structured and operated. What is clear, however, is that the USA is willing to create an interim stockpile for the bank and that its aim is to prevent proliferation and not to promote the use of nuclear power.

V. Discussions in INFCE about a nuclear fuel bank

The task of Working Group 3 of INFCE was to study the problem of "assurances of long-term supply of technology, fuel and heavy water and services in the interest of national needs consistent with non-proliferation". This obviously covers, among other things, the relevant fuel-supply assurances of the NNPA.

During discussions on the fuel bank, it was pointed out that existing contractual arrangements for supplies had generally worked satisfactorily in the past in terms of security of supply and that they would probably continue to do so in the future. However, greater certainty and predictability with regard to national import, export and non-proliferation policies would enhance market efficiency—governmental intervention had caused delay and expense. Moreover, the efforts of some countries to apply updated non-proliferation conditions had had the consequence of increasing the perception of vulnerability to supply interruptions of countries dependent on nuclear imports. To date, no agreements had included explicit commitments by governments not to interfere with nuclear exports or imports under existing contracts. The resulting concern over possible interruptions could also impede the achievement of an effective non-proliferation régime by encouraging consumers to use interruptions in the issue of import or export licences to seek to achieve a lowering of non-proliferation restraints on existing contracts or to seek autarky in the fuel cycle.

Agreement was reached in the Working Group that the main and preferred mechanism for assured supply was a competitive market, stockpiling and diversification. However, since developing countries were less able to protect themselves against risks of disruption of supplies, the need for an alternative mechanism was generally recognized.

Three alternative mechanisms were suggested during discussions:

1. Back-up or emergency arrangements making joint use of existing or additional stockpiles, which, according to the Working Group, might prove expensive and could destabilize the market.

2. An international fuel bank as an insurance mechanism for countries with small nuclear power programmes, to provide supplies in the event of a contract default that is not the result of a violation of a non-proliferation commitment; the assets of the bank would be (*a*) fissionable materials transferred by gift or consignment to an international stockpile controlled by the bank, (*b*) material earmarked for sale and retained within the jurisdiction of the supplier member, (*c*) commitments to provide enrichment and other services to be used in the specified emergency circumstances, and (*d*) cash contributions. Relevant materials would be subject to IAEA safeguards. Membership access conditions, disposition of spent fuel, prices and the financing and structure of the bank were also discussed.

3. Multinational facilities or fuel cycle centres which, on the one hand, would, in themselves, presumably not be able to reduce automatically the risk of interruption of supply, and which, on the other hand, would have to incorporate adequate controls and adopt stringent rules in order to give the world community confidence that their services would be used for peaceful purposes only.

The special needs of developing countries were discussed in relation to long-term supply assurance. Even if, as was pointed out, the situation in individual developing countries differs widely, common to most of them is the initial small scale of their nuclear programmes and their scarce resources. This would indicate that the topics of discussion here, and particularly the fuel bank, may be of relevance to them, since despite the greater needs of the developing countries compared with those of the industrialized countries, the present institutional arrangements do not provide any additional assurances.

VI. Concluding comments

There is no doubt that a congruence exists between the international nuclear fuel cycle pool as suggested by Yugoslavia in 1976 and the outcome of discussion in the INFCE Working Group 3 on the subject of the fuel bank. The IAEA Statute provides a basic institutional mechanism which could accommodate both of these ideas. On the other hand, the points of departures and motivation behind INFCE's and Yugoslavia's ideas are quite different.

The outcome of the INFCE Working Group 3 discussions related to the fuel bank bears the imprint of current US policy by the mere fact that INFCE itself was initiated and influenced by Carter's 1979 pronouncement. Present US policy and the 1978 NNPA reflect the school of thought "which introduces the new element of 'temptation' and holds that the widespread dispersion of sensitive facilities and materials implicit in earlier policies would create grave risks of diversion and proliferation even on the part of countries with no premeditated design to go nuclear" (Goldschmidt & Kratzer, 1978). As such, both the Act and US policy, which were being imposed on other countries, are in general regarded as discriminatory and unacceptable. It is felt that the fuel bank envisaged by the Act will serve only the purpose of relieving the difficulties resulting almost exclusively from the Act itself.

While the NNPA is motivated primarily by desires for non-proliferation, with the proposed nuclear fuel bank as an instrument of the non-proliferation policy of the nuclear weapon states, the Yugoslav suggestion is motivated primarily by interest in the promotion of nuclear power in accordance with Article IV of the NPT. Here the nuclear fuel cycle pool, or several pools formed by different groups of countries, are envisaged as serving mainly the interests of the developing countries based on the assumption that the non-military use of nuclear energy will benefit, and is the responsibility of, every country in the world.

The fact that nuclear proliferation is rapidly spreading is not the fault of the non-nuclear weapon states, nor is it due to the increase in nuclear power production facilities outside IAEA safeguards.

The advantages of a pool over a bank

Restraints on nuclear technology are part of a broader problem imposing the need for the introduction of a new world economic order. The pool should contribute, basically, to healing the problem at its roots by promoting economic progress in the less developed world, nuclear power being part of such progress. The international fuel cycle pool as proposed by Yugoslavia is different from the international fuel bank approach in two basic ways: (*a*) the motivation behind it is promotion of peaceful nuclear power under safeguards conditions, along the lines of the Colombo and Havana resolutions, rather than non-proliferation, and (*b*) while the international fuel bank is intended as an emergency mechanism, the fuel cycle pool aims at assisting the developing countries in providing their fuel cycle needs regularly and developing their own material and human resources in accordance with their economic interests and power, without discrimination. Since it would be based on different motives, it would have an advantage over a bank in that it would require and stimulate the active technology and political involvement of its member states and, in addition, would place full responsibility for non-proliferation in their own hands, thereby creating a system of mutual safeguards. A wide base for securing the non-proliferation of nuclear weapons would thereby be created, in contrast with the philosophy of the past which mainly leaves the responsibility for safeguards in the hands of the great powers. To bring the national responsibility of pool members for non-proliferation into full effect, the national systems of accounting for and control of nuclear materials would have to be given full importance and co-operation established to form a consistent international framework of safeguards, strengthening the assurances of peaceful use of nuclear energy. The membership of the pool would have to be based on equal rights and responsibilities in administering the financial and material assets of the pool. While observing democratic principles, the organization would have to achieve full functionality and efficiency of operation. Here it should be emphasized that the IAEA should use its professional abilities to elaborate on the many aspects of the pooling of resources by the developing and other countries. The role of the IAEA would be strengthened in view of the many contributions it would make to the various aspects of the creation and operation of such a pool.

Despite possible differences in motivation behind the concepts of a bank and a pool, it is always possible to find ways of satisfying the common interest if there is one. In this case, a common interest, that is, peace, certainly does exist. Current political developments and the communications explosion have meant that the democratic approach cannot be limited to some parts of international life only. Nothing can remain secret, and no benefits can be denied to any nation. Based on such principles, the international nuclear fuel cycle pool, as a means of promoting the peaceful uses of nuclear power without discrimination, can usefully be established to serve the majority of humanity. However, while these principles are clear and simple, to engineer a working international agreement on them, based on international consensus, may be difficult and time-consuming.

References

Goldschmidt, B. & Kratzer, M., 1978. *Peaceful nuclear relations: A Study of the creation and the erosion of confidence* (ICGNE, New York and London).

Paper 13. An international fuel bank

D.L. SIAZON, Jr*

Embassy of the Philippines, Peter Jordan Strasse 37, A-1109 Vienna 19, Austria

I. Introduction

The Non-Proliferation Treaty (NPT) rests on the basic bargain that non-nuclear weapon states party to the Treaty would be guaranteed free access to equipment, materials, and scientific and technological information for the peaceful uses of nuclear energy in exchange for an undertaking not to manufacture, acquire or develop a nuclear weapon or a nuclear explosive device. As an integral part of this bargain, the developing states party to this Treaty were guaranteed special considerations with respect to nuclear supply, and the nuclear weapon states have undertaken to pursue negotiations in good faith towards an early cessation of the nuclear arms race.

Two factors are clearly weakening the non-proliferation effects of the Treaty: (*a*) continuing vertical proliferation, and (*b*) the imposition (by a group of supplier countries) of new non-proliferation measures as conditions for nuclear supply. With regard to (*a*), it should be borne in mind that horizontal non-proliferation measures can succeed in the long run only if vertical nuclear proliferation is stopped. Regarding (*b*), unilaterally imposed supply conditions have seriously undermined the guarantee of free access to nuclear supply granted to non-nuclear weapon states party to the Treaty, thus affecting the balance of the basic trade-off. These measures suggest to the non-nuclear weapon states that the development of their endogenous nuclear capability is the only way to avoid unpredictable supply conditions and supplier controls intrusive of their national sovereignty.

Fortunately, however, partly as a result of the International Nuclear Fuel Cycle Evaluation (INFCE), there is a growing realization by major supplier countries that onerous and unilaterally imposed conditions for nuclear supply will most probably help to accelerate rather than delay

*The opinions expressed in this paper are those of the author and do not necessarily reflect the official policy of his professional affiliation.

nuclear proliferation (Campbell, 1979; Bettauer, 1978). There is a realization that, to restore the fragile balance in the NPT's basic trade-off, effective and practicable measures are needed, serving to improve the consumer countries' perceptions of the reliability of nuclear supply. One such mechanism is an international nuclear fuel bank that would guarantee the provision of services—related to the front end of the nuclear fuel cycle—to all countries that qualify as recipients.

It has been argued, however, that such a bank would not be necessary because the commercial or free market system has worked very well in the past. Utility companies have successfully protected themselves from aberrations in the nuclear supply system by the traditional methods of stockpiling, purchases on the spot markets, diversification of sources, or swap arrangements with other utilities. While utility companies in industrialized countries, relying solely on commercial markets, have experienced little difficulty in dealing with problems related to nuclear supply, it cannot be denied that the developing countries and industrialized countries with small nuclear power programmes could face serious operational difficulties by relying solely on commercial markets.

The non-proliferation undertakings in the NPT are government undertakings, and the assurances of nuclear supply are likewise government assurances. The mechanism for assuring nuclear supply must therefore transcend the normal commercial markets and involve government guarantees. This is the essence of the Treaty's basic compromise.

The developing countries party to the NPT have special rights under Article IV.2, rights that have been confirmed by all parties to the Treaty. The establishment of an international nuclear fuel bank would be a means of ensuring the exercise of these rights. Some industrialized non-nuclear weapon states party to the Treaty may decide not to seek government assurances for nuclear supply. They might consider themselves capable of dealing with varied nuclear export procedures by relying solely on the commercial markets. The developing countries are, however, not in the same happy situation. Their limited resources and small nuclear power programmes compel them to acquire a more secure and stable medium for nuclear supply. This can be met by an international nuclear fuel bank.

II. Basic elements of an international nuclear fuel bank

The basic elements of an international nuclear fuel bank will have to be included in an agreement to be negotiated among interested states. In a way, this process of negotiation has started in the course of the work of Working Group 3 of INFCE. With the termination of INFCE in early 1980, work will have to be continued elsewhere. In order to facilitate the conclusion of an agreement, some elements are being suggested for consideration in the establishment of an interntional nuclear fuel bank.

Membership

The rationale for establishing an international nuclear fuel bank derives from the rights and obligations of nuclear and non-nuclear weapon states as set down in the NPT; consequently, the condition for access by a state to the bank's services should be the application of the International Atomic Energy Agency (IAEA) safeguards to all its peaceful nuclear activities. In addition, such a state would have to pledge not to manufacture, develop or acquire a nuclear weapon or a nuclear explosive device. States in this category could be known as consumer members.

The conditions for states wishing to provide services to the bank should be sufficiently flexible to enable the optimum number of donor and supplier countries to participate, thus contributing towards the effectiveness of the bank and the strengthening of the non-proliferation régime. States in this category could be called supplier members. It should be possible for a supplier member simultaneously to be a consumer member, provided the state also meets all the conditions for consumer members.

Conditions of provision of services

Any consumer member should be able to acquire services from the bank:

(*a*) when it has been unable to obtain materials or services from any supplier within a specified time after the requirement for such materials or services and after having exerted earnest efforts to do so for at least one year, provided such failure was not due to non-acceptance or violation of non-proliferation conditions internationally agreed upon by both supplier and consumer countries, or

(*b*) when it has been unable to obtain materials or services under a contract with a supplier within a specified period after the stipulated date of delivery, provided the consumer member has complied with the terms and conditions of the supply contract and all applicable international agreements it has made concerning the peaceful uses of nuclear energy.

The condition referred to in paragraph (*a*) above is important to developing countries that may find themselves suddenly unable to secure needed materials or services because of *force majeure* or exceptionally adverse commercial conditions, which have unexpectedly evolved after the start of the relevant nuclear project. In a sense, this is like a reversed scheme for the stabilization of export earnings. Article IV.2 of the NPT envisages the need for special consideration to be given to developing countries party to the Treaty, and the conditions for service in paragraph (*a*) clearly fall within that specific provision. Even the Uranium Institute, which has not been sympathetic to the establishment of an international nuclear fuel bank, has in an official paper argued: "An international guarantee fund, preferably managed by the International Monetary Fund (IMF), could be made available to utilities, via their governments, in credit-tight developing countries" (Uranium Institute, 1979).

The use of bank services under paragraph (*a*) above may be considered technical assistance and be limited to developing countries who are members of the bank. In this connection, it is to be noted with regret that according to IAEA statistics, technical assistance provided by the IAEA for the period 1974–78 amounted to only US $45.7 million or an average of $9 million per year distributed among about 70 developing countries.

The condition in paragraph (*b*) above may apply to any non-nuclear weapon state qualifying as a consumer member. This would protect consumer member states from the adverse effects of unilateral changes in non-proliferation conditions by a supplier that was supposed to provide materials or services under existing contract. At the same time, this would insulate consumer members from the economic burdens of *forces majeures* in the supplier country.

There may be serious objections to the condition in paragraph (*b*) because of the large resources the bank would require to provide a dependable insurance scheme. In addition, this may be interpreted as possibly leading to serious interferences in the commercial markets; therefore, it may be useful to limit the bank's services to developing countries and those other countries that have very small nuclear power programmes. With more limited clientele, fewer adjustments in the commercial markets would be required and fewer resources would be necessary.

The conditions for access to the bank's services should not be limited to cases where inability to secure materials and services, stipulated in existing and valid contracts, is due to an imposition by a supplier of controls not provided for in the supply contract. Such a situation might encourage suppliers to change supply conditions, because they would be in a position to rationalize that the consumer country would not suffer from such action. While this may be true in certain cases, frequent changes in supply conditions may lead to an unwarranted escalation of non-proliferation conditions. This might be perceived by consumer members as a convincing argument that stability of nuclear supply can be assured only by the development of endogenous nuclear fuel capabilities, a situation which could contribute to faster nuclear proliferation.

The details for the provision of materials and services through the bank will have to be negotiated and clearly spelled out in the agreement establishing the bank. Procedures will have to be clear, and a certain degree of automaticity will have to be introduced in the provision of services to consumer members. Consumer members would have to pay for such services at the time of delivery; however, the pricing policy of the bank would have to take into account Article IV.2 of the NPT.

Financing

The administrative costs of the bank should be borne through assessed contributions of both supplier and consumer member states, in accordance with a scale of assessment based on that of the United Nations. The administrative costs would include necessary expenses for staff, offices and facilities required for the effective functioning of the bank.

The costs of materials and services made available through the bank and not covered by payments of consumer members should be borne by the supplier countries. The costs of safeguarding the materials and services belonging to or provided through the bank should be borne in accordance with the provisions of the IAEA Statute.

Assets

The natural low-enriched uranium, heavy water and capital that would be made available to the bank for its effective operation would be considered the bank's assets and be located in a place to be determined by the member countries. The bank's assets should be maintained at a level sufficient to ensure reliability of supply. Supplier members could provide the bank's assets through outright contributions of cash or kind, the consignment of materials and services, or the earmarking of materials and services. In order to ensure the stability of the bank's resources, materials or services donated, consigned or earmarked by a supplier member to the bank should not be permitted to be withdrawn unless the supplier member were prepared to contribute sufficient cash to cover the purchases of such materials or services at the time of the withdrawal of the consignment or of the earmarking. In addition, the transfer and disposition of materials secured through the bank should be made in accordance with conditions and procedures established beforehand by the bank and should not be subject to additional conditions and/or procedures required by the supplier state concerned. The bank should, however, ensure that its nuclear materials are under safeguards and that the disposition of irradiated special nuclear material produced through the use of materials or services supplied through the bank should be in accordance with procedures previously determined by the bank.

Structure of the bank

The bank should have a General Council or a General Conference, composed of all members of the bank. This council should meet at least once a year, or oftener, upon the request of a majority of members. The council would have a limited role in the actual operations of the bank: it would be responsible for the approval of the annual budget, the election of the

Executive Head of the bank, the determination of the headquarters of the bank, and the provision of resources necessary to enable the Executive Head to operate the bank effectively. The council should not have the power to suspend or expel members except in cases where members had ceased to meet the conditions of membership as specified in the agreement establishing the bank. This would insulate the bank's operation from the vagaries of international politics.

The Executive Head would be responsible for the appointment of the members of the Secretariat and for the operational decisions necessary for the effective functioning of the bank. He would prepare the budget and report on a regular basis to the council on the activities of the bank. The term of the Executive Head would be for four years with the possibility of one re-election.

Entry into force and duration

The agreement establishing the bank should enter into force as soon as a certain number of supplier states and consumer states became members. The duration of the agreement should be 15 years, so that it would expire at the same time as the NPT, and should be subject to extension by consensus of all members. A withdrawal by a supplier member should take effect only after two years from the date of receipt of notification of that supplier's intention to withdraw. Amendments to the bank's agreement should be possible only by a very difficult procedure, preferably by consensus of all members.

Relations with the IAEA

The IAEA is an established organization with proven competence in the promotion of the peaceful uses of nuclear energy. In considering an international nuclear fuel bank, the possibility of using the existing IAEA institution must be examined. Article 9 of the IAEA Statute provides for the possibility of the IAEA assuming the envisaged functions of a fuel bank. While there are distinct advantages in using the IAEA as a fuel bank, there are also some significant disadvantages which favour the establishment of a new organization.

If the bank were to be within the IAEA machinery, the Board of Governors would have to exercise control over the bank's functions. This might raise certain difficulties because political problems in the Board would inevitably spill over into the bank's activities. This development would weaken the reliability of the bank's assurances of supply. At the same time, if the bank were within the IAEA, conditions for membership in the bank could not be legally different from conditions for IAEA membership. The bank's services would have to be made available to all

IAEA members. This might run contrary to the purposes of the bank.

In view of these reasons, it seems preferable to have an international fuel bank separate from the IAEA, but periodically furnishing the IAEA with reports of its activities.

III. Conclusions

The world is presently engaged in a search for additional, effective non-proliferation measures. There is an earnest desire to build on the success achieved by the NPT and the better understanding emanating from INFCE. If the world is to succeed in this endeavour, it must convince those countries that have the greatest incentive to breach the non-proliferation barrier that there are more benefits to be attained by observing the non-proliferation policy. The establishment of an effective international nuclear fuel bank is a step in that direction.

References

Bettauer, R.J., 1978. The Nuclear Non-Proliferation Act of 1978, *Law and Policy in International Business,* 10 (4).

Campbell, R., 1979. Basic concepts for nuclear commerce. Statement delivered at the American Nuclear Society Conference on International Nuclear Commerce, New Orleans, Louisiana, USA, 9–11 September.

Uranium Institute, 1979. The nuclear fuel bank issue as seen by uranium producers and consumers. Circulated at the American Nuclear Society Conference on International Nuclear Commerce, New Orleans, Louisiana, USA, 9–11 September.

Paper 14. International plutonium storage

M.L. JAMES*

Plutonium Management Programme, International Atomic Energy Agency, Vienna International Center, A-1400 Vienna, Austria

I. Introduction

There is no single sweeping answer to the problem of nuclear proliferation. The approach must continue to be, as it has always been, a range of measures interlocking, relating and supporting each other. International plutonium storage (IPS) is just one of the measures which have the potential to strengthen the non-proliferation régime.

The concept of IPS was incorporated into the Statute of the International Atomic Energy Agency (IAEA), in Article XII.A.5, when it was founded 23 years ago. Its aim is the physical international control of plutonium at the most sensitive stage of the fuel cycle—the storage and handling of plutonium in separated form after reprocessing and before use.

While controls on materials cannot be a fundamental solution to the proliferation problem, they have the value of offering reassurances against the diversion of material from civil to military use, acting as a direct disincentive to diversion and as a reassurance against the *fear* of diversion by another state or states. So far the IAEA has concentrated, in its safeguards programme, on one kind of control—detection of diversion. IPS would offer an additional level of security by aiming at the prevention of diversion at a particularly sensitive stage of the fuel cycle. The distinction between prevention and detection is real, but only experience will show how significant it is. So far, the IAEA safeguards system has not reported a diversion, and although there is a high probability of diagnosis, we are uncertain about the cure. On the other hand, during the time that weapons-usable material would be under international control, the danger of its clandestine diversion should not arise, and that of overt diversion should certainly be reduced.

---

* The opinions expressed in this paper are those of the author and do not necessarily reflect the official policy of his professional affiliation.

II. International plutonium storage

IPS would reinforce and complement safeguards on reactors, reprocessing plants and fuel fabrication. The placing of separated plutonium in *internationally* controlled stores would still the fears of those who claim that safeguards alone cannot offer sufficient reassurance that plutonium in *national* stores—stored in significant quantities for long periods and in a readily weapons-usable form—is proof against diversion.

IPS should not be considered in isolation but in relation to safeguards on plutonium during various stages of the fuel cycle, such as: (*a*) in reactors (in which plutonium is present only in irradiated fuel and therefore highly inaccessible), (*b*) during reprocessing (while there is a level of protection by radioactivity, the plutonium remains inaccessible), and (*c*) during fuel fabrication (where plutonium is more accessible). The relative accessibility of plutonium in different stages of the fuel cycle is suggested in the table in Appendix 1. International control under IPS would be a measure added to safeguards at the particularly attractive stage for diversion when plutonium is separated and is not protected by serious radiation or containment in a production process.

IPS and safeguards, taken together, would therefore aim to produce a combined régime which would reassure the world community that all civil plutonium is either (*a*) in properly safeguarded use in reactors, reprocessing or fabrication plants, or research, or (*b*) in internationally controlled storage. In the case of (*a*) there would be a high degree of probability that diversion would be detected promptly, and in the case of (*b*) diversion would involve open confrontation with an international authority.

Why plutonium

There is a tendency to focus on plutonium as it has direct use in both power generation and nuclear weapons. This is understandable, but the potential for the use of enrichment technology to produce highly enriched uranium should also be viewed seriously. Enriched uranium is an equally effective material for nuclear explosives. Recent suggestions that particular states might be seeking to develop a weapon capacity (whether true or false) have, indeed, generally been based on the assumption that the enrichment route was being followed. In the total context of non-proliferation one must, therefore, give equal attention to enrichment and plutonium separation as sources of weapons-usable material. This paper is, however, confined to plutonium.

Stockpiling of plutonium

The stockpiling of plutonium is not a theoretical future problem, as is sometimes implied by opponents of reprocessing. Stocks have been growing for some years and have already reached large quantities. Further growth is inevitable.

The danger of diversion of plutonium is not primarily a quantitative problem: about 8 kg is a significant quantity. Nonetheless, it is worth noting that in 1978, 19 tonnes of separated plutonium were under safeguards. Information supplied by states (excluding those with centrally planned economies) to Working Group 4 of the International Nuclear Fuel Cycle Evaluation (INFCE)—which is the most up-to-date data we have—projects that by 1990 a further 126 tonnes of plutonium will have been separated and that by the year 2000 this will have increased to over 400 tonnes. How much of this projected separation would be in stock rather than in use at those dates is speculative, but the information accumulated by Working Group 4 suggests at least 25 per cent would be in stock on both dates.

Whatever the outcome of the present international focus on non-proliferation, particularly in INFCE, it is a fact that plutonium has been separated for many years in the civil fuel cycle and that significant stocks of it already exist in national stores, including stores in non-nuclear weapon states. A number of states adopted, from the beginning, a fuel cycle including reprocessing, and it is clear that this policy will generally continue, whether reprocessing is seen primarily as a way of disposing of irradiated fuel or as a source of fuel for recycle, especially in fast breeder reactors. Moreover, since the plutonium fuel cycle has the unique quality of being able to create more fissile material than it consumes, it would be unrealistic not to accept that fuel cycles using plutonium will be developed increasingly.

It is also inevitable, in states working towards fast breeder reactor programmes, that in at least the next 20 or 30 years the amounts of separated plutonium will not be matched by immediate requirements. It will be stockpiled, either under national or international auspices, in preparation for the large initial plutonium requirements of fast breeder reactors. It is therefore desirable to find a way of controlling the use of plutonium so that the legitimate aspirations of states to develop fuel cycles using it, particularly in fast breeder reactors, can be realized without unnecessary impediments, while offering reassurance against a further spread of plutonium-based weapons.

III. The IAEA plutonium management programme

The exploratory study

In 1976, the Director General of the IAEA initiated a study on the possibility of introducing a system of international plutonium storage in accordance with Article XII.A.5 of the IAEA Statute. A report was circulated to all member states in July 1978 (IAEA, 1978). It aroused considerable interest, and, as a result, in December 1978 an Expert Group was convened to prepare a detailed scheme. The Group consists of experts from all member states (see table 1) wishing to take part and is attended by representatives of 24 countries and the Commission of the European Communities. The Expert Group represents most of the states who might be directly concerned in a scheme—both Third World and developed countries, and suppliers and customers. It represents a wide cross-section of interests and stages of nuclear development.

Table 1. Delegations which took part in the International Atomic Energy Agency Expert Group on International Plutonium Storage[a]

Argentina	Mexico
Australia	Netherlands
Belgium	Pakistan
Brazil	Poland
Canada	Spain
Denmark	Sweden
Egypt	Switzerland
Finland	USSR
France	United Kingdom
Germany, Federal Republic of	USA
India	Yugoslavia
Italy	
Japan	Commission of the European Communities

Note:

[a] Any member state of the IAEA may attend the Expert Group. The Nuclear Energy Agency of the OECD (Organisation for Economic Co-operation and Development) has been invited to attend future meetings.

Expert Group on international plutonium storage

The task of the IAEA group was to prepare proposals for a scheme for submission to the IAEA Board of Governors. A factual, but personal, account of its deliberations is given below.[1]

The starting-point for the Expert Group was Article XII.A.5 of the IAEA Statute, which provides for: (*a*) deposit with the Agency of plutonium separated in member states but excess to their immediate requirements for peaceful and safeguarded uses in reactors or research, and (*b*)

[1] The author is secretary of the IAEA Expert Group on International Plutonium Storage and responsible to all its members. He would, however, stress that he is not speaking for the group collectively. It is only possible for the 25 separate delegations and their governments to state their individual points of view.

prompt release by the Agency of deposited plutonium, under continuing safeguards, for peaceful use in reactors or research. This Article of the Statute does not specify institutional and administrative arrangements for its implementation, and it is with these that the Working Group is dealing.

Non-proliferation and security of supply

It became clear early that, in looking at possible controls on plutonium, it was necessary to take account of two basic considerations: (*a*) non-proliferation of nuclear weapons, and (*b*) security of supply of materials for the nuclear industry as an essential energy source. In devising a scheme which would be acceptable to all the states concerned, whether suppliers, reprocessors or users of plutonium, both factors were equally important.

The international climate in which the Expert Group started work was one in which some of the recent measures taken by states, in the implementation of their non-proliferation policies, had led to difficulties and fears concerning the assured and stable supply of materials and equipment—fears, therefore, about the fundamental security of energy supplies. This had added to the prevailing international atmosphere of doubt, uncertainty and mutual mistrust, with a negative rather than a positive effect on international relations.

It was clear that the Expert Group would have to devise proposals for a scheme which would neither threaten, nor appear to threaten, national programmes for the peaceful use of nuclear energy, such as plutonium recycle in thermal or fast breeder reactors and related research and development. Indeed, given the obstacles to such developments, which have arisen from national supply policies, it was desirable that a scheme should make properly safeguarded and peaceful work somewhat easier to pursue.

Non-discrimination and national sovereignty

The Expert Group was also reminded that the security of energy supply was seen as something basic to national security, particularly after the energy problems encountered in the last few years. Policy for energy control was seen as an aspect of national sovereignty, and no scheme perceived as threatening that sovereignty would be viable. In more concrete terms, this meant that states were not willing to place a supra-national body—such as the IAEA Board of Governors or the administration of an IPS scheme—in a position to decide whether states required plutonium for their nuclear industry. That was a decision for the states themselves.

However, states expressed a willingness to accept controls which demonstrated that plutonium produced by, or supplied to, them was genuinely employed in peaceful energy uses: that is, to accept that from this point of view their plutonium operations should be 'transparent' so that others might be reassured as to their peaceful intentions. Transparency implies a readiness to accept a system of notification, checking and controls which would make it possible for states collectively to know which states held separated plutonium, and to be reassured that plutonium was either in verified and peaceful uses in reactors or research, or if not in such uses, that plutonium was being stored under international control. Thus an acceptable approach appeared likely to confine a scheme strictly to generally agreed non-proliferation objectives.

In addition, much stress was laid, in the Expert Group's discussions, on non-discrimination. Some have seen this as a backwash from reactions to the Nuclear Suppliers Group guidelines by states which were not party to them. It was clear that an acceptable scheme should not discriminate between nuclear weapon states and non-nuclear weapon states or between resource-supplying and resource-consuming countries.

The basic requirements

The Expert Group therefore started work in the knowledge that, to be generally acceptable to the states concerned, a scheme would have to: (*a*) strengthen the international non-proliferation régime in a credible way, which would reassure the world community against the spread of plutonium-based weapons; (*b*) not discriminate between states; (*c*) not interfere with national energy programmes, and (*d*) reduce the obstacles in the way of the development of plutonium fuel cycles, as a contribution to the general development of peaceful nuclear energy.

IV. Outline of a scheme

Registration and deposit of plutonium

Starting from this basis, the Expert Group is considering a scheme on the following lines. It should be borne in mind that the views of states still differ on a number of points.

It was proposed that, in the interest of transparency of plutonium operations, all plutonium separated in states taking part in an IPS scheme should be registered with the international body controlling the scheme.

On registration, plutonium in excess of the state's immediate requirements would then be deposited in internationally controlled storage. Many members of the Expert Group appeared to agree that the simplest approach would be to regard as excess any plutonium which had not been authorized for release prior to registration. Some delegations had reservations on this and would like to find another formula.

General undertakings of states

It has been envisaged that in addition to being required both to register all plutonium separated and deposit excess plutonium in this way, states taking part in a scheme might also give certain general undertakings on joining a scheme in relation to plutonium which would later be released to them. Among the undertakings which have been suggested are: (*a*) to use released plutonium, either directly or in research, only for peaceful purposes; (*b*) to place all plutonium within the state's jurisdiction under IAEA safeguards, and (*c*) to provide physical protection for released plutonium in accordance with the recommendations published by the IAEA from time to time (such recommendations are found in INFCIRC/225/Rev.1).

Internationally controlled storage

The Expert Group has given considerable attention to the location and operation of international stores in more detail than can be reflected in this paper.

It seems generally agreed that it would be impractical, and unnecessarily costly, for the IAEA to build and operate a network of stores in the relevant countries, although the possibility of an internationally managed store in one or two locations has not been ruled out altogether. The general view is that internationally controlled storage should be effected at places where plutonium would be stored in quantity, in any event, as a result of normal industrial operations: notably at the output of reprocessing plants and at the input to fuel-fabrication plants. In practical terms, this would mean the establishment of international storage on a fairly limited number of sites in the foreseeable future.

This approach would minimize the possibility of IPS affecting the smooth flow of normal nuclear industrial operations. It would also not increase the transport of plutonium above what would normally be required by the industry without an IPS scheme in operation. This is desirable in view of the security risks, and costs, associated with plutonium transport.

The thinking of the Expert Group is, therefore, that excess plutonium should be stored in the operator's stores, which would be required in any case at reprocessing and fuel fabrication plants.[2] The deposit and release of plutonium would, however, be controlled by international officers stationed permanently at the stores. These officers would take custody of plutonium when it was deposited and would not release it except on the authorization of the international body responsible for the scheme.

In addition to these officers having legal control of deposit and release, it has been proposed that, to increase the security and credibility of a scheme, they should also have physical control by being the joint keyholders

[2] One delegation would prefer to confine international storage to reprocessing plants.

of the storage area within the storage installation on a two-key basis with the operator of the co-located plant (the inner storage area is marked 'Pu STORAGE' in figure 1). This would make it physically impossible for plutonium to leave the store legitimately without the agreement of the international authority. A technical advisory group is advising the main Expert Group on possible arrangements for operation on this basis.

Figure 1. Diagram of main features of a plutonium store

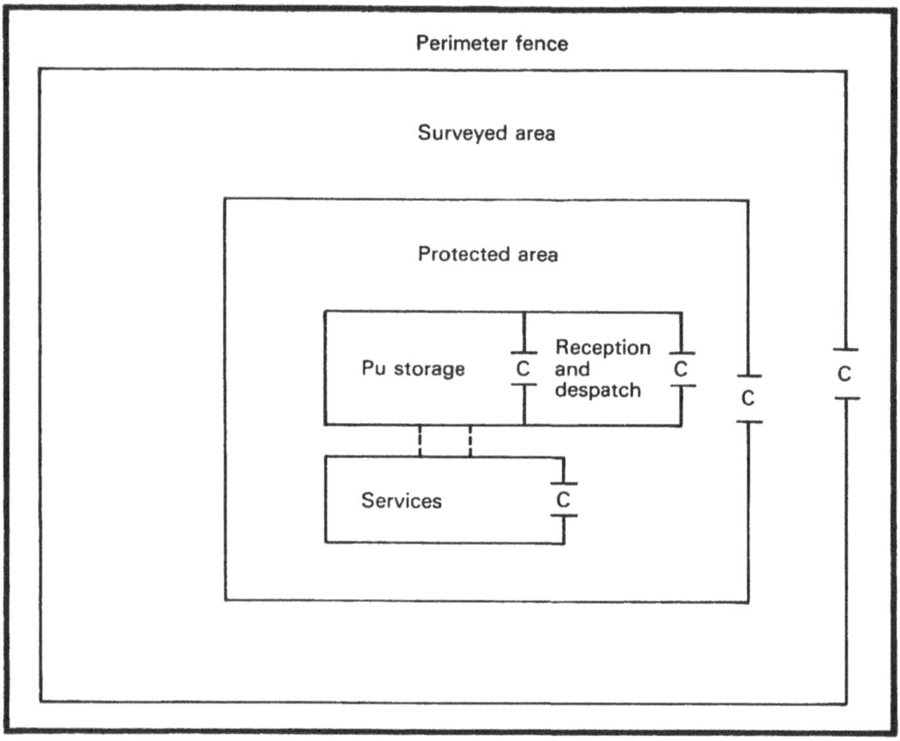

Notes:

C = controlled doors, which may also have fissile material detectors, etc. Physical protection measures are not shown in the figure.

It is envisaged that the store would be operated under a host state agreement with the state on the territory of which it stood. In such an agreement the state would undertake not to interfere in the international control of deposit and release at the store and to provide adequate physical protection. The routine aspects of store operation would remain in the hands of the operator: provision of ventilation and cooling, monitoring of conditions, application of the host state's health and safety regulations, and so on.

Release of plutonium

Once plutonium has been deposited, it is envisaged that a state would have the right to its prompt release on request, for uses that satisfied the criteria of Article XII.A.5, that is, uses that were: (*a*) peaceful, (*b*) under IAEA safeguards, and (*c*) specified. Criterion (*a*) might be covered by a general undertaking, such as that given above under *General undertakings of states.* Criterion (*b*) would require an appropriate safeguards agreement, and (*c*) would make it possible to check the authenticity of the proposed use and the appropriateness of the quantity and timing of releases in accordance with an agreed authorization procedure.

Most of the plutonium would clearly be released for reactor fuel. It is agreed that the physical despatch of plutonium from international storage should not take place until the material can be taken immediately into the proposed use. This should minimize the time that plutonium would remain in an accessible separated form after release. The intention would be for it to be fabricated into fuel and placed in a reactor, where it would become radioactive and inaccessible, as rapidly as possible.

It is envisaged that the procedure for authorization of release would be fairly simple. A state might submit an application giving details of: (*a*) the requirement—quantity of plutonium and date needed; and (*b*) the intended use—for example, fabrication of MOX fuel in plant A for reactor B. So far, (*b*) has been seen as a fairly detailed statement which would cover, *inter alia,* in the case of reactor use, the capacity of the fuel-fabrication plant, the date fabrication would start, the length of time it would take, the point at which the fuel would be loaded into the reactor, the fuel composition and the capacity of the reactor.

This statement of the facts that had given rise to the requirement would be checked for factual accuracy, on the basis of verified information supplied by the state and by inspection of facilities. Release would be authorized provided that: (*a*) the material was to be covered by IAEA safeguards, and (*b*) the checking procedure showed that the plutonium would be taken into use immediately or the quantity requested would be absorbed rapidly by the proposed use.

The IPS administration would have no power to pass judgement on the need of the state for the proposed use and therefore would have no discretion to refuse release if the simple requirements, outlined above, were met.

It is envisaged that there would be a follow-up, co-ordinated with safeguards operations, to verify that the material was, in fact, used for the purpose for which it had been released. It has also been suggested that if material were not taken into use within a specified period, there might be a requirement for its return to international storage.

To simplify the operation of the procedure, it has been proposed that once a scheme was operating, states might usually submit applications for programmes of releases, some time, perhaps years, ahead of the physical need for the plutonium. Application might, in fact, be made in appropriate cases before the plutonium was separated by reprocessing. Thus, decisions might be well ahead of physical need (which would avoid the danger of hindering industrial operations), but plutonium would remain in international custody until the agreed release date. Equally, the procedure would be designed so that it could be completed quickly in the case of short-term needs.

While it might be possible for decisions to release plutonium to be taken by authorized officers, it is the view of most of the Working Group that any decision to refuse release should only be taken collectively by representatives of all the states party to the scheme and then only on the grounds that the simple release criteria had not been met. In the event of such a refusal, the requesting state would have the right to refer the matter to a form of arbitration specified in the instrument establishing the scheme and adhered to by all states on joining it.

V. Conclusions

There appears to be a genuine desire to reach agreement on an IPS scheme. If this can be done, it will be a significant step in developing the world non-proliferation régime. Such a scheme should also be beneficial, in the context of easing problems related to the supply of plutonium, in that it would increase reassurance against the possible misuse of it and thus lessen the present burden of multiplying controls. Whatever the result, it will be an indication of what is practicable in international nuclear relations in the post-INFCE period.

Reference

IAEA (International Atomic Energy Agency), 1978. *The International Management & Storage of Plutonium and Spent Fuel*, IAEA Report.

Appendix A

Alternative forms of plutonium at different points in the fuel cycle

Table A.1 (reproduced by courtesy of the Technical Secretariat of INFCE Working Group 4) indicates by reference letters A to G the decreasing time and resources needed to separate directly usable material for nuclear weapons from the gamma-active and other protective elements, at different points in the fuel cycle. A indicates the lowest level of protection, and G the highest. The table is illustrated in figure A.1.

Figure A.1.

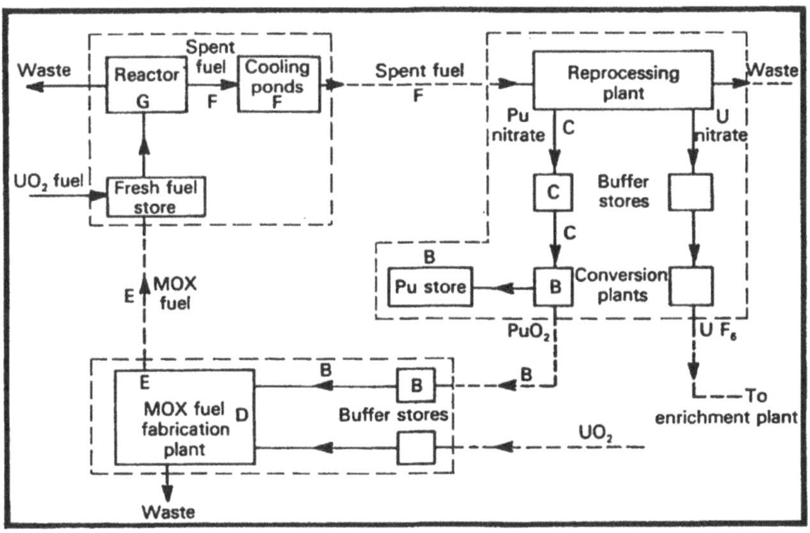

For Table A.1. see overleaf

Table A.1. Alternative forms of plutonium at different points in the fuel cycle

Level of protection	Material classification (IAEA)	Material	Location in fuel cycle (incl. transportation between facilities)	Processing necessary	Conversion time to metallic components of nuclear explosives (IAEA)[1]	Approx. mass needed for one significant quantity of Pu[1]	Inherent barriers to misuse
G	3/4	Fuel assembly during irradiation	Within reactor core	Impossible	—	∽ 1 tonne (2 PWR sub-assemblies)	Intensely radioactive
F	3	Discharged irradiated fuel sub-assembly	Reactor storage, ponds, interim store, reprocessing plant	Mechanical and chemical separation followed by conversion	Order of months (1—3)	∽ 1 tonne (2 PWR sub-assemblies)	Intense radio-activity falling with time after discharge
E	2	Fuel sub-assembly (prior to irradiation)	Recycle fuel fabrication plant, reactor site	Mechanical and chemical separation followed by conversion	Order of weeks[3] (1—3)	∽ 140 kg (1 PWR sub-assembly) ∽ 50 kg (1 LMFBR sub-assembly)	Toxicity, radioactivity[2]
D	2	$(Pu + U)O_2$	Recycle fuel fabrication plant (possibly reprocessing plant)	Dissolution and separation followed by reduction to metal	Order of weeks[3] (1—3)	∽ 140 kg[4] (PWR fuel) 50 kg (LMFBR fuel)	Toxicity, radioactivity[2]
C	2	$Pu(NO_3)_4$	Reprocessing plant	Conversion to fluoride followed by reduction	Order of weeks[3] (1—3)	∽ 30 kg @ 300 g/litre	Toxicity, radioactivity[2]
B	2	PuO_2	Reprocessing plant, plutonium storage site, recycle fuel fabrication plant	Conversion to fluoride followed by reduction to metal and fabrication	Order of weeks[3] (1—3)	∽ 10 kg	Toxicity, radioactivity[2]
A	1	Pu metal	Does not appear in fuel cycle	Fabrication only	Order of days (7—10)	8 kg	Toxicity, radioactivity[2]

[1] These items are still under discussion by the IAEA.

[2] Depends on the plutonium isotopic composition.

[3] While no single factor is completely responsible for the indicated range of 1—3 weeks for conversion of these Pu and U components, the pure compounds will tend to be at the lower end of the range and the mixtures and scrap at the higher end.

[4] Will vary as MOX composition varies, for example, of "masterblend"

Paper 15. Institutional solutions to the proliferation risks of plutonium

J. LIND*

International Nuclear Affairs and Non-proliferation, Ministry for Foreign Affairs, Box 16121, S-103 23 Stockholm, Sweden

I. Introduction

Nuclear weapon proliferation is a political problem rather than a technical one. But as long as political measures do not achieve success, other means are necessary for preventing the proliferation of nuclear weapons. Current efforts towards non-proliferation include a vast number of technical and institutional measures related to peaceful nuclear energy programmes. Further international attention to such measures will need to focus on the strengthening of already existing arrangements and on possible new institutional measures at the international level.

In discussing new international measures that could be implemented in regard to the management of plutonium, this paper places particular emphasis on various steps which could enhance international confidence that plutonium will not be used for other than peaceful, non-explosive purposes. Management of spent fuel and separated plutonium in national energy programmes must be undertaken in a safe, economic and predictable manner, which are all important interrelated considerations, for such international confidence to be gained.

II. The case of plutonium

Plutonium is a by-product of energy production based on nuclear fission of uranium-235. Operating reactors are continuously producing this nuclear material which exists in all spent fuel.

* The opinions expressed in this paper are those of the author and do not necessarily reflect the official policy of his professional affiliation.

Plutonium can be used as fuel, either in thermal or in fast breeder reactors, as well as in research. Research would normally involve only small quantities. Thermal recycle has taken place only on a very limited scale, and for a number of reasons the prospects for it in the near future seem highly uncertain. Further, initial fuelling of fast breeder reactors will be a proposition in a very few countries and only in the fairly distant future. For these reasons, in the foreseeable future very significant portions of all plutonium produced in peaceful, non-explosive nuclear energy programmes will be in excess of projected use. Storage of plutonium will thus be necessary, some in separated form. Most of the excess quantities, however, appear likely to remain in spent fuel for the time being. The International Atomic Energy Agency (IAEA) study on international management and storage of spent fuel should be seen against this background.

Lack of confidence

In separated form plutonium can be utilized more or less directly in nuclear explosives. Timely detection of any diversion of a significant quantity, which is the fundamental objective of IAEA safeguards, appears difficult to ensure where this weapons-usable nuclear material is concerned. In other words, the international community cannot be confident that IAEA safeguards, as presently conceived, can prevent the abuse of separated plutonium. This is why questions of reprocessing, the use of reactor-produced plutonium and plutonium storage have created so much international concern and have been the object of increasing attention in the context of the non-proliferation of nuclear weapons. There seems to be growing international consensus that the characteristics of plutonium warrant such attention. Much less unanimity and even a considerable amount of controversy have, however, been expressed as to which new measures would be best, in the sense of both effectiveness and international acceptability, in order to minimize the proliferation risks. Various technological as well as institutional measures involving plutonium have been discussed. Deferral of reprocessing, thermal recycling, co-processing, spiking, co-location of reprocessing and MOX-fuel fabrication are all examples of technological measures. Regional nuclear fuel cycle centres, multinational reprocessing or storage facilities and international ownership, management or storage of spent fuel or separated plutonium are examples of institutional measures.

III. Closing the plutonium credibility gap

Joint objectives

A discussion of possible new measures, by which IAEA safeguards might be supplemented in relation to separated plutonium, must also take into account what appears to be a growing resentment of some existing non-proliferation measures relating to supplier–recipient relationships. These have sometimes been accused of being discriminatory or disruptive to the planning and implementation of national energy programmes, or both. In the search for new measures, the avoidance of any further criticism of this nature must be a goal equal in importance to that of enhancing international confidence that no abuse of separated plutonium will take place. Thus, any approach aimed at closing what might be termed the plutonium credibility gap would really need to strike a balance between: (*a*) enhanced international confidence that a given state will not divert or otherwise abuse separated plutonium, and (*b*) enhanced national confidence that there would be tangible returns, as regards assurances of supply, for states accepting non-proliferation measures.

These two points would usefully serve as joint objectives for any new measures to be considered. International agreement by all involved that both these objectives should be pursued would certainly be an extremely important step in the direction of solving what has become an important international problem. Efforts within INFCE (the International Nuclear Fuel Cycle Evaluation), and, more specifically, in the IAEA Expert Group on International Plutonium Storage have made important contributions in the direction of establishing consensus on such joint objectives.

Building confidence

The pursuit of the greatest possible international acceptance of the above two objectives is probably the fundamental point of departure in the work to identify, discuss and, hopefully, agree on appropriate new non-proliferation measures. Even with considerable progress in the direction of establishing international consensus on these objectives, further efforts aimed at agreed specific measures to achieve them will be highly complex. Various approaches towards such efforts may be considered: (*a*) to start by defining an optimum mix of technical and institutional means by which the political objectives could be reached—the task then would be to assess all the specific measures which are required in order to arrive at that mix, or (*b*) to seek agreement one step at a time. Each step would be helpful in its own right and would create constructive prerequisites for further steps. Such a step-by-step approach to the problem will be considered below.

First Step

A possible first step along such an approach would be to define agreed uses and other management practices relating to plutonium. At least to begin with, it might be sufficient to consider the possibility of co-ordinated unilateral assurances given by states involved. A development of this would be an international agreement on permissible uses of plutonium. Such assurances, whether in unilateral or other form, could cover all forms of plutonium including such quantities as are contained in spent fuel. Alternatively the assurances could be limited to separated plutonium. Specific assurances would probably be necessary in order to meet the basic objectives. The term 'specific assurances' implies rather narrowly defined uses or other management practices as well as the provision of fairly detailed information on quantities, timing, exact facilities involved and so on.

Agreed uses and other management practices would in one way or another involve: (*a*) storage under conditions to be defined; (*b*) use as fresh fuel in fast breeder reactors and possibly as recycled material in thermal reactors, and (*c*) research for peaceful, non-explosive purposes.

Details as to quantities, timing, and so on, related to these different possible uses of plutonium, would need to be analysed in detail in order to arrive at objective and generally acceptable solutions.

As noted above, use of plutonium as fuel or in research would, in relation to total amounts generated, involve only small quantities. Thus, questions relating to storage will necessarily have to be given considerable attention. This is presently being done by the IAEA. Agreement on problems of procedures and principles for release from storage certainly would be made considerably easier should it prove possible to arrive at agreement on uses and other management practices. An international understanding on such management practices for plutonium would also be, most probably, a helpful step towards the joint objectives of enhanced confidence mentioned above.

However, even a detailed scheme of agreed uses containing specific assurances could possibly raise questions, for example, as regards verification. Such questions would imply a need for further steps.

Further Steps

An important step towards the verification of agreed uses would be for the state involved to announce in some appropriate form the specific whereabouts and usage of separated plutonium for peaceful purposes and possibly of all plutonium contained in spent fuel. End-use statements of this nature could be made to other states which had accepted the same procedure. Under a more developed scheme, statements might be made to the IAEA or some other international body instituted specifically for the purpose. The countries or the international body receiving such statements would be able to conclude fairly easily whether the stated specifics for storage or use coincided with the agreed uses.

An added element of confidence could be achieved if, in addition, agreement could be reached that traditional IAEA safeguards be expanded in relation to separated plutonium so that declared storage or use could be verified by international officers. This would probably necessitate registering all plutonium in the nuclear fuel cycle with the IAEA and adding a number of procedures to material accountancy practices presently involved in safeguards.

If implementation of the steps discussed above still did not appear to fulfil the joint objectives of confidence, the next step would probably include more detailed institutional and procedural means. Besides IAEA safeguards, these could provide for plutonium, in use or storage, to be either placed in international custody or at least subject to detailed international controls. New international advisory or deciding bodies would then probably have to be instituted. The definition of their specific role and areas of competence would require most careful international efforts. The resulting scheme would probably deserve to be designated as an International Plutonium Management Régime.

IV. Conclusions

Plutonium, especially when in separated form, causes special concern from a nuclear weapon proliferation point of view. There seem to be growing international awareness and acceptance that this concern is warranted and that some international measures are called for in addition to the IAEA safeguards and other existing arrangements. A further strengthening of this awareness would be a significant contribution. At least equally important, however, would be international agreement as to the basic objectives to be pursued. The joint objectives proposed in this paper can be summarized as the promotion of confidence concerning: (*a*) diversion of separated plutonium for explosive purposes, and (*b*) the supply of nuclear materials.

A step-by-step procedure for the consideration and, hopefully, implementation of such measures could involve the following: (*a*) agreed uses and other management practices; (*b*) end-use statements; (*c*) verification and end-use, and (*d*) an International Plutonium Management Régime. Each of these possible steps would be likely to contribute in its own right to enhanced confidence in the joint sense indicated by this paper. The steps would be mutually supportive. Agreement on one would facilitate implementation of the following steps. At the same time, the greater the consensus attained on specific objectives, the greater would the prospects be for progress in the political context of non-proliferation. The need for additional technical and institutional measures would be likely to diminish accordingly.

Finally, agreed objectives and steps relating to plutonium could be an interesting basis for the further search for such political, technical and institutional measures involving the entire nuclear fuel cycle as might deserve the designation 'new nuclear non-proliferation régime'.

Paper 16. International storage of spent reactor fuel elements

B. GUSTAFSSON*

Svensk Kärnbränsleförsörjning AB, Box 5864, S-102 48 Stockholm 5, Sweden

I. Introduction

This paper deals with the management of light water reactor (LWR) spent fuel during interim storage, after its discharge from the reactor and up to the time when it is either reprocessed or sent to a repository for disposal after appropriate conditioning. Spent fuel generation, storage technologies, transport, investment costs and finally some national approaches to storage of spent fuel will be discussed.

II. Spent fuel generation

Spent fuel generation figures are presented in several documents, for

Table 1. Projected quantities of accumulated LWR spent fuel in the USA and Western Europe

Year	USA (tonnes uranium)	Western Europe (tonnes uranium)
1980	3 000	3 300
1985	13 000	12 800
1990	30 000	30 000
2000	77 000	80 000

Sources:

Nuclear Assurance Corporation (1979); US Nuclear Regulatory Commission (1979).

*The opinions expressed in this paper are those of the author and do not necessarily reflect the official policy of his professional affiliation.

example, from the OECD (Organisation for Economic Co-operation and Development) and the International Nuclear Fuel Cycle Evaluation (INFCE). Table 1 presents projections of the quantities of LWR fuel that may be accumulated in the USA and Western Europe. Whatever future spent fuel policies will be, significant amounts of spent fuel will need to be stored, and there will be a shortage of storage capacity if no actions are taken.

III. Design bases of existing technology

Until now the general approach to the management of spent fuel has been to store it in fuel storage pools integrated with the power reactor units. Fuel discharges of boiling water reactors (BWRs) are presently approximately one-fourth to one-fifth of the core a year and for pressurized water reactors (PWRs), approximately one-third of the core a year. In most countries and for most of the LWR power stations in operation or under construction, a total capacity of one and one-half core loads for BWRs and one and two-thirds core loads for PWRs is provided for the storage of spent fuel elements in the storage pools. The above-mentioned storage capacities are a licensing requirement in some countries.

At-reactor (AR) storage

All LWR power stations have a spent fuel storage area as an integrated part of the plant. The reasons for storing spent fuel at the power stations are: (*a*) to provide an interim capacity between refuelling operations and the transportation link; (*b*) to provide the capacity for temporarily storing a full-core discharge in case of a complete reactor or fuel-element inspection, and (*c*) to obtain sufficient cooling and decay time before transport of the spent fuel. The capacity at each power station is determined by discharge rate, core load and necessary cooling time.

Away-from-reactor (AFR) storage

In the case of AFR facilities spent fuel is stored in a structure not integral to a reactor, which may be co-located with another nuclear facility or may be on a separate site. Presently such facilities exist only at reprocessing plants. Examples of such facilities in the United States are: (*a*) Nuclear Fuel Service (NFS), West Valley; (*b*) General Electric (GE), Morris, and (*c*) Allied Gulf Nuclear Service (AGNS), Barnwell. Examples in Western Europe are: (*a*) British Nuclear Fuel Limited (BNFL), Windscale; (*b*) Compagnie Générale des Matières Nucléaires (COGEMA), La Hague; Wiederaufarbeitungsanlage von Kernbrennstoffen (WAK), Karlsruhe, and (*c*) European Company for Reprocessing of Irradiated Fuels (Eurochemic), Mol. The capacity of these AFR facilities varies from 100 to more than 1 000 tonnes uranium.

Usually, in the case of AR and AFR facilities, spent LWR fuel is stored under controlled conditions in water-filled pools. The water serves to shield personnel from radiation from the stored fuel and is also used to remove heat generated by radioactive decay. The water is continuously purified in order to remove both fission and corrosion products.

An AFR facility is elaborate in its design, compared with an AR facility. This is mainly because of: (*a*) various types of spent fuel to be handled—from PWRs and BWRs; (*b*) transport cask reception, cask handling, maintenance and so on, and (*c*) the larger amounts of spent fuel to be stored. Wet storage of LWR spent fuel is today a well-proven technology on an industrial scale. So far, with the experience of storage times of up to 20 years, no particular problems have been encountered.

IV. Storage technologies under development

Several studies are presently under way in order to improve present technology and to develop new ones. The greatest effort has so far been directed towards LWR fuel.

AR facilities

The most common approach for AR facilities has been to install compact racks. The use of compact racks can involve the incorporation of a neutron-poison absorption material in a stainless-steel matrix, thus allowing a much smaller centre-to-centre spacing between the fuel elements. Such arrangements may increase the storage capacity for discharged fuel at an AR facility by a factor of two to three. This storage technique is being adopted at many AR facilities, and several licences are presently being handled.

AFR facilities

Wet-storage facilities are presently the most common and proven, worldwide. However, several types of dry-storage facility are now under development. The main efforts have so far been made in FR Germany and the USA. Dry storage of LWR spent fuel has, however, not yet been applied on a large scale.

The various types of dry-storage facility under development are listed below.

1. A retrievable surface-storage facility (RSSF) with an air-cooled vault: the spent fuel is encapsulated in some kind of material, such as carbon steel. The package is placed in a concrete vault located on ground level and is cooled by forced air circulation through the vault.

2. An RSSF with a shielded, sealed cask: the spent fuel is encapsulated, as in the former type, and placed in a concrete cylinder which acts as a gamma and neutron shield. The package is placed above ground, and cooling is achieved by natural air convection. These two types were originally developed for interim storage of solidified high-level waste and are now being considered in the USA for interim storage of spent LWR fuel.

3. Dry-caisson storage: this involves the utilization of the shielding and heat-transfer qualities of the earth. The fuel elements are encapsulated in mild steel and placed in a well caisson. The depth of such a caisson is in the range of 7–8 m in order to provide adequate shielding.

4. Air-cooled storage racks: the spent fuel is stored in a reinforced concrete building located under ground and is probably placed in racks without encapsulation. Cooling of the spent fuel is obtained by forced air circulation.

5. Dry storage in transport casks: in FR Germany and the USA different facilities involving dry storage in transport casks are under development. These casks are designed in accordance with the International Atomic Energy Agency (IAEA) type B-U transport licensing requirements. The spent fuel would be loaded into the casks at the power station and then intermediately stored at the reactor site or transported to another site which would then serve as an AFR facility. Transport casks of this type (heavy-wall structure) can most probably be stored in conventional buildings, even in countries that have to consider a number of external impact events because of licence requirements.

Dry-storage facilities have some advantages compared with wet storage, such as: (*a*) less maintenance, (*b*) possible cooling by natural convection, (*c*) reduced amount of radioactive waste, and (*d*) no radiolysis because of the absence of water.

On the other hand there are disadvantages relating to dry storage, and some problems exist.

1. Due to the high decay heat, the spent fuel must be stored in water pools until encapsulation and transfer to dry interim storage.

2. Additional, complicated process facilities are needed because of the required encapsulation procedure.

3. In some of the designs, high-efficiency filtration systems which would cause large amounts of radioactive waste are called for.

4. There is a much higher fuel temperature in dry storage than in wet storage, and the long-term integrity of the fuel-element cladding and structural material is not well known or proven.

5. As the described facilities would be only for intermediate storage, the spent fuel would, after a certain period of time, have to be transferred elsewhere for further treatment. Because of the uncertainty concerning the integrity of the cladding and structure material, further handling of the spent fuel might cause unexpected difficulties.

6. At the moment no licensing procedures and guidelines for dry-storage facilities exist.

V. National approaches

The USA

The US Department of Energy (DOE) is making plans in order to fulfil the storage requirements of Mr. Carter's spent fuel policy of 18 October 1977. The DOE study on AFR storage concludes that spent fuel storage capacity will be a concern in the early 1990s regardless of the growth of nuclear power. Different alternatives have been studied, such as: (*a*) the utilization of existing storage technologies to increase AR storage capacity and the free use of storage space in facilities integrated with reactors; (*b*) the trans-shipment of spent fuel freely from facilities with full pools to pools with available storage capacity within each utility-owned reactor system, regardless of geographic location, and (*c*) the complete and free interchange of storage space regardless of ownership or geographic location.

Applying alternative (*a*), additional AFR capacity would be needed as early as 1980. The need is expected to grow to 1 900 tonnes of uranium by 1985 and to reach 27 000 tonnes of uranium by the year 2000. With alternative (*b*), the need for additional AFRs would not occur until 1982. The requirements in 1985 and in the year 2000 are expected to be 700 tonnes and 19 300 tonnes, respectively. Alternative (*c*) would postpone any AFR requirements until 1999.

In order to cope with the spent fuel management problem in the USA in the coming years, a combination of the first and second alternatives described will have to be adopted until a future spent fuel policy is decided upon.

FR Germany

FR Germany is planning an integrated fuel cycle centre to be in operation sometime between the late 1980s and early 1990s. As an intermediate solution, AR storage capacity is being increased by means of compact racks. In addition, contracts for reprocessing of spent fuel have been signed with

France, and an AFR storage facility for 1 500 tonnes of uranium is planned to be in operation by 1985.

The licence application for this AFR—a wet storage facility to be located at Aahaus—was submitted by the Deutsche Gesellschaft für Wiederaufarbeitung von Kernbrennstoffen (DWK) in 1978 (Baatz, Essman & Janberg, 1979). DWK is about to supplement its former licence application for wet storage with another application for dry storage in transport storage casks, also at Aahaus.

According to DWK, the investment cost is lower for dry transport-storage cask facilities than for wet storage. This is mainly because of the very severe West German safety regulations (protection against aircraft crashes and earthquakes, and high redundancy requirements).

France

France has opted for reprocessing. Consequently the AFR facilities are integrated parts of the reprocessing facilities. Existing AFR facilities, as well as planned ones, involve water-cooled storage pools. Currently there is one spent fuel storage pool for LWR fuel at the La Hague reprocessing facility, with a capacity of approximately 350 tonnes of uranium. Additional pools are under construction, each module with a capacity of approximately 1 000 tonnes. Completion of these pools is scheduled for the period 1981–84.

Sweden

The magnitude of spent fuel generation depends on how many reactors will be put into operation during the coming years. Assuming that 12 power reactors will be in operation during the second half of the 1980s, approximately 2 300 tonnes of uranium would be accumulated by 1990. Contracts for reprocessing spent fuel discharged during the 1970s and 1980s are in force, in total covering 810 tonnes of uranium. The quantity of spent fuel which is not contracted for reprocessing is planned to be stored in a central spent fuel storage facility (CLAB) to become operative in late 1984.

Since 1977, the Swedish Nuclear Fuel Supply Co. (SKBF) has been conducting a project aiming to obtain licences for and to design, erect and operate this storage facility. Start of construction is planned for 1 May 1980, and commissioning of the facility is scheduled for late 1984. Three different sites were considered during the conceptual study. Permission to locate the CLAB facility on the Simpevarp site was given by the Swedish government on 14 December 1978.

As the CLAB facility is, today, the only licensed independent LWR-AFR facility in the Western World, it might be of interest to describe the licensing procedure. Table 2 shows the required permits, along with their respective application and approval dates.

Table 2. CLAB licensing procedure

Permit	Submitted	Approved
Site permit according to building act § 136a	30 Nov 1977	14 Dec 1978
Application according to the Environmental Protection Act	29 March 1979	10 Jul 1979
Application according to the Nuclear Energy Act	30 Nov 1977	23 Aug 1979[a]
City plan for the Simpevarp site	12 Feb 1974	15 Mar 1979
Exemption from obligation to apply for building permit	21 Mar 1979	24 Aug 1979

[a] Recommendations on 29 June 1979 from the Swedish nuclear inspectorate to the government for approval according to the Nuclear Energy Act.

The Swedish government approval according to the Nuclear Energy Act was granted on 23 August 1979 on condition that construction work on the site not be started before 1 May 1980. This condition was imposed due to the planned advisory nuclear referendum, which was to take place in early 1980.

The CLAB design, shown in figure 1, allows for a storage capacity of 3 000 tonnes of uranium with a potential for increasing the capacity to 9 000 tonnes. The facility will also be able to store reactor core components. A 3 000-tonne capacity equals 12 years' discharge from a 12-reactor domestic system.

Figure 1. Central spent fuel storage facility (CLAB)

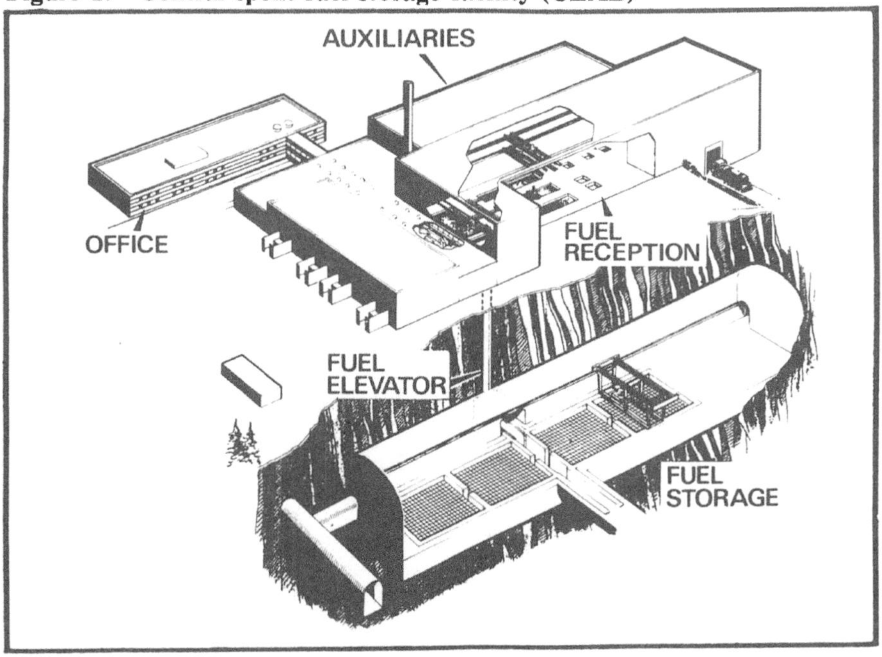

The four water-filled storage pools will be situated under ground in a cavern with a rock cover of approximately 25 m. Each pool will have a capacity of 750 tonnes. The dimensions of the cavern housing the storage pools will be 100 m × 21 m × 27 m. The fuel reception and auxiliary buildings will be located at ground level.

VI. Investment costs

Investment costs for AFR facilities vary greatly depending on local economic conditions, national licensing requirements and the size and type of facility chosen. Investment costs published so far have varied between US $50 and $200 per kilogram of uranium. The investment for the 3 000-tonne uranium CLAB facility is calculated (in 1979 prices, not adjusted for inflation) at approximately US $250 million which corresponds to approximately US $80–$85 per kilogram of uranium.

VII. Transport

An AFR storage facility, such as CLAB, can be successfully operated only if reliable and safe transport arrangements can be secured.

Because of the geographic location of the Swedish nuclear power stations and the CLAB facility, sea transport of spent fuel is planned. For this purpose SKBF, together with a Swedish consultant, has since early 1979 been planning the 'roll-on roll-off' ship shown in figure 2. The main

Figure 2. 'Roll-on Roll-off' spent fuel transport ship

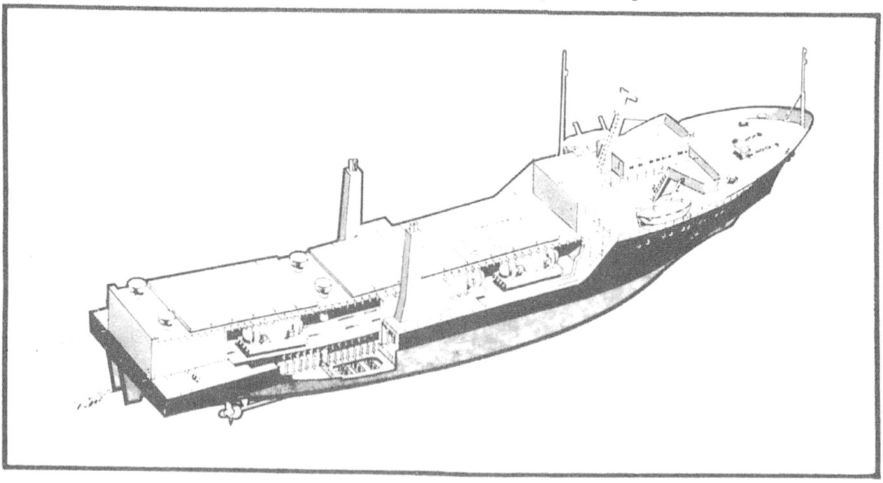

requirements of the design are: (*a*) that the boat will remain afloat after serious damage and that the machinery and electrical systems will have high damage resistance (IMCO class I rules); (*b*) effective fire-fighting facilities; (*c*) a ventilation system in the cargo hold for cooling purposes; (*d*) reliability as regards seaworthiness, manoeuvrability and navigability in ice (it must meet Baltic Ice class 1A requirements); (*e*) radiation shielding for living and working areas; (*f*) instrumentation for radiation measurement; (*g*) good navigation and communications equipment, and (*h*) special means in order to facilitate the search for and salvage of a sunken vessel.

According to the present time schedule, the ship will start operation in 1982.

VIII. International management and storage of spent fuel

In the coming years it is likely that the provision of additional spent fuel storage capacity will generally be handled by the expansion of existing national-facility capacities and by the provision of national AFR facilities. It is appropriate to raise the question whether or not, in the long term, multinational or international co-operation in spent fuel management could offer advantages over strictly national approaches. Storage and management of spent fuel under international co-operation will, however, depend greatly on the willingness of individual countries to offer sites and accept the agreed international conditions which could have precedence over national laws. The issue of public acceptance of foreign spent fuel for storage within a country would be of equal importance.

The role which the IAEA and other international organizations may play was discussed in INFCE and is being examined in the IAEA study presently in progress on international management and storage of spent fuel.

References

Baatz, H., Essmann, J. & Janberg, K.G., 1979. Uranium Institute, Fourth Annual Symposium in London, 10–12 September.

Nuclear Assurance Corporation, 1979. *Light Water Reactor Spent Fuel Storage and Reprocessing in Europe,* Report, January.

US Nuclear Regulatory Commission, 1979. *Final Generic Environmental Impact Statement on Handling and Storage of Spent Light Water Power Reactor Fuel,* Report Nureg-0575, Vols 1–3, August.

Paper 17. Spent fuel storage

G. I. ROCHLIN*

Institute of Governmental Studies, University of California, Berkeley, California 94720, USA

I. Introduction

The need to take simultaneous account of spent fuel as an energy resource, a potential weapon material and a long-lived hazardous waste is the basis for present difficulties in determining how it should be managed, by what means, and in what places.

If the potential resource value of the contained plutonium were to be forgone, spent fuel itself could be encapsulated and treated as a waste. Thus, waste disposal is not in itself a persuasive argument for reprocessing (OECD, 1977).

There is, however, a manifest reluctance to plan for disposal of spent fuel. Most countries would prefer to have their spent fuel reprocessed, either to recover the energy or economic value or as an investment in future fast breeder reactor (FBR) development. But spent fuel storage will be required in any case, since reprocessing capacity will not be adequate even for then-current arisings until at least the mid-1990s (OECD, 1977).

The question is not whether spent fuel storage will exist, but why it has attracted so much international attention. Clearly it is because some fuel suppliers have been promoting it as an alternative to reprocessing for the time being. The political salience of this otherwise rather trivial technical problem arises directly from such non-proliferation-related concerns, although nuclear exporters are, to a lesser extent, also interested in reducing the accessibility of spent fuel as the source material for plutonium.

International management, oversight or control of spent fuel have thus emerged both as a fuel cycle issue in itself and as an element of suggestions for the internationalization of plutonium stocks (IAEA, 1978). On the surface, this appears to be a relatively straightforward problem, particularly if states can retain title to the stored fuel or rights for eventual return of an

*The opinions expressed in this paper are those of the author and do not necessarily reflect the official policy of his professional affiliation.

equivalent amount of plutonium, or if they can be issued cash or fuel-equivalent credits for the surrendering of title. There are, however, several questions that need to be answered before it can be determined what forms of internationalization, if any, would satisfy a broad enough range of opinion to have a reasonable chance of being negotiated. Among these questions are:

1. Can spent fuel be stored safely, and if so, for how long?
2. What is the risk of diversion or theft of spent fuel and how can it best be minimized?
3. Is present national capacity for spent fuel storage adequate, and if not, can it be increased?
4. How homogeneous are attitudes and policies of the developed countries, and how do they compare to the attitudes of Third World countries?
5. Are fuel customers willing to store their spent fuel domestically if permission to reprocess is denied, and if not, what can be done?
6. Are fuel suppliers willing to take back spent fuel to prevent its reprocessing or storage under national control, and if not, what other provisions can they make?

These questions will be discussed in turn.

II. Can spent fuel be stored safely?

The discussion here will centre on spent fuel from light water reactor (LWR) or Canadian heavy water (CANDU) designs. Although these differ in initial enrichment and in the fraction of fission products, plutonium and other transuranics contained in spent fuel, they have in common the general form of the fuel, sintered uranium oxide, and the use of a zirconium–aluminium alloy (zircalloy) for cladding. The great bulk of commercial spent fuel over the next two decades will be of this form. British AGR (advanced gas-cooled reactor) fuel is stainless-steel clad, but otherwise similar. Older gas–graphite and Magnox design reactors have fuel with less-stable cladding that could present problems for design of extended storage facilities, but such reactors will be few and located in countries that have reprocessing facilities for them already.

At the moment of reactor shut-down, spent LWR fuel generates about six per cent of the heat generated at full reactor power. Most of this energy, however, is released by the decay of very short-lived isotopes. After only 10 days, the thermal power drops by a factor of 20. The spent fuel is transferred during this time to an at-reactor water-cooled storage pool for further cooling. After about 120 days, most of the iodine-131, an isotope particularly hazardous to life owing to its high activity and accumulation by the thyroid, has decayed away, and the thermal power has also dropped by a factor of 100. Spent fuel could be shipped off-site any time after this, although cooling must be maintained for some time (see Paper 16).

A major element in the safety debate has been the integrity of the fuel cladding (Windscale Inquiry, 1978). Even small leaks could release gases such as krypton-85. But experience with storage of high burn-up, zircalloy-clad LWR fuel for more than a decade has as yet shown no actual or incipient clad failures (Johnson, 1977). On this basis, spent fuel could be stored safely for at least one or two decades without further treatment, although continuous inspection and monitoring would certainly be warranted in view of the relatively small amount of experience. Storage for longer periods, or by methods that make continuous inspection difficult, might require that spent fuel bundles be sealed into special canisters or containers filled with a heat-conducting substance such as helium gas or a soft metal. This would provide greater assurance without much increasing the difficulty of later recovery.

Design of the storage facility itself is not technically difficult, and there are many possibilities (see Paper 16). Table 1 contains a list of spent fuel storage alternatives at present being considered by the USA. The choice to be made in any country will depend upon the assumptions made as to the length of storage, the degree of protection to be afforded against accident or external disturbance, and the cost.

Despite uncertainties about what choices will be made, the emerging general consensus is that spent LWR fuel can be stored safely for at least one or two decades (and possibly much longer) and that no more than a small fraction will be so damaged that it must be disposed of rather than retrieved for later reprocessing, should that course be selected (DOE, 1979*a*). If packed in containers, it can be stored retrievably for many decades.

III. What is the risk of diversion or theft?

When discharged from the reactor, spent LWR fuel is so radioactive that fully shielded, remote handling of bundles is required. Although the radiation barrier decreases with time, as shown in table 2, it is still high enough after 30 years that theft by non-state adversaries is quite unlikely. Moreover, LWR spent fuel bundles are physically quite large (several hundreds of kilograms apiece) and would have to be transported in rather massive containers.[1] These barriers would, of course, present few

[1] Fuel assemblies for a pressurized-water reactor weigh 550–650 kg apiece. Those for a boiling water reactor weigh 250–300 kg.

Table 1. Comparison of spent fuel storage alternatives[a]

Storage alternative	Confinement barriers[b]	Means of heat removal	Method of controlling corrosion	Maintenance requirements	Surface land use
Unpackaged storage					
Water basin	Water and filters	Forced circulation of basin water	Water-quality control	High	Low
Air-cooled vault	Filters	Forced circulation of air	Low temperature	Moderate	Moderate
Packaged storage					
Water basin	Water, filters, canister	Forced circulation of basin water	Packaged in inert or non-corrosive medium	High	Low
Air-cooled vault	Canister	Natural circulation of air	Packaged in inert or non-corrosive medium	Low	Moderate
Concrete surface silo	Canister	Natural circulation of or conduction to air	Packaged in inert or non-corrosive medium	Low	High[c]
Geologic (deep subsurface)	Canister, hole liner, filters	Conduction to earth	Packaged in inert or non-corrosive medium	Moderate	Low
Near-surface caisson	Canister, hole liner	Conduction to earth	Packaged in inert or non-corrosive medium	Low	High

[a] Adapted from DOE (1979*b*), table 5.7.1, Volume 3.

[b] In addition to fuel cladding.

[c] Estimated to be 120 000 m^3 for 1 500 tonnes, assuming one tonne/storage unit.

Table 2. The radiation barrier of a representative spent pressurized water reactor fuel assembly

Time since discharge	Radiation flux at midplane[a] (rads/h at 1 m)
At discharge	1 500 000
1 month	170 000
150 days	35 000
1 year	12 000
5 years	2 000
10 years	1 400
30 years	1 200

[a] Assumed to be from a current design 1 150 MW(e) PWR with burn-up of 33 000 MW-days/tonne. Radiation levels from spent boiling-water reactor fuel elements will be a factor of two to three lower due to lower burn-up and physically smaller size.

Source:

Miller (1978).

problems to national groups, whether clandestine or open, with access to standard fuel-handling equipment.

This has led to fears that extended spent fuel storage would increase proliferation risks by creating plutonium 'mines' (Marshall, 1978). However, the incremental risk seems small. Most countries with operating reactors will have spent fuel up to five-years-old stored, as a matter of course, at reactors. But as shown in table 2, the radiation barrier decreases by only 40 per cent between 5 and 30 years, and hardly at all between 10 and 30 years. The technology needed for reprocessing would be only marginally simpler, if at all, after many decades (Hannerz & Segerberg, 1979). It is not that the risk of the diversion of an individual bundle increases with time, but rather that national diversion from a large storage facility may become more difficult to detect as inventories increase. Fuel from over 100 reactor-years of operation may be in store, while the plutonium from a single year's discharge of one LWR is enough to make several nuclear explosives.

The problems of dealing with spent CANDU fuel are somewhat different. Fuel bundles are smaller, more numerous and have a lower net radioactivity. The plutonium in CANDU fuel is closer to traditional weapons-grade than is that in LWR fuel. On the other hand, the net plutonium inventory per individual bundle is smaller. On balance, the risks of spent CANDU fuel appear to be slightly higher for national diversion and about the same for theft by subnational groups.

In sum, the radiation barrier of spent fuel appears to be sufficiently high that adequate protection against subnational theft can be provided with a modicum of physical security. Protection against diversion or seizure by the host state may be more difficult. The implementation of near real-time monitoring would provide an added level of safeguards against diversion, as would multinational or international control of the sites.[2] The risks of outright seizure can be reduced by storage on foreign soil, but the risk reduction is not great unless spent fuel is shipped early enough to prevent the accumulation of inventories at the reactors themselves.

IV. Is present national capacity adequate?

At present, spent fuel is being stored primarily at reactor sites. Some limited additional capacity is available, but this is at reprocessing plant sites (see Paper 16). Total capacity is limited at present, and many reactor operators will soon be faced with full pools and nowhere to which to ship the excess.

At-reactor (AR) capacity could be expanded by a variety of means, most of which are technically straightforward. The alternative is the construction of new away-from-reactor (AFR) facilities. These can be wet or dry, above ground or below, with encapsulation or without it (see Paper 16).

None of these measures is technically or industrially very complicated. Present national capacity for storage is inadequate primarily because it was widely assumed until only a few years ago that spent fuel could soon be reprocessed (OECD, 1977). If it remains inadequate, that will be due to a combination of the following political constraints, quite idiosyncratic to each country (*a*) difficulty in resolving the AR versus AFR debate; (*b*) environmental objections to extended spent fuel storage; (*c*) legal restrictions related to waste management requirements for licensing, and (*d*) unwillingness to accept domestic storage as an alternative to reprocessing or retransfer to another location.

V. Present attitudes and policies[3]

Among the states most advanced in nuclear power development (Belgium, France, FR Germany, Japan, the UK, the USA and the USSR), maintenance of the power industry is the leading issue. Neither France, the UK nor, presumably, the USSR has a storage problem at the moment. Moreover, the UK is considering extended water-cooled storage of

---

[2] Real-time monitoring is continuous, for instance via a monitored data link or TV channel. Near real-time monitoring involves the use of recording cameras or data accumulators that can be periodically interrogated at time intervals which are short compared with the time needed to effect a successful diversion or theft.

[3] For a more complete region-by-region analysis of spent fuel storage capabilities and the prospects for multinational or international storage see Williams & Deese (1979).

solidified high-level reprocessing wastes, which is easily adaptable to spent fuel if necessary.

The USA is currently in the midst of an intense internal argument as to whether AFR storage is needed, and if so, when. This debate is further coloured by the internal dispute over reprocessing policy and waste-repository timing, and by the Congressional debate over the proposed US acceptance of limited amounts of spent fuel from abroad. In the wake of the Gorleben decision, FR Germany is only now beginning an internal debate over AFR versus AR storage. Japan is unwilling to accept either, and intends to continue shipping spent fuel to France and the UK for reprocessing under contract.

Among the other countries with more advanced nuclear technology—the CMEA (Council for Mutual Economic Assistance) countries, other Euratom members, Finland, India, South Korea, Sweden, Switzerland and Taiwan—only India is reprocessing or has immediate plans. CMEA countries have fuel return agreements with the USSR, as has Finland for Loviisa-1. For the rest, spent fuel is presently more a waste that needs management than a present resource. All these countries are capable of expanded AR or AFR storage, if necessary, but would much prefer to have spent fuel reprocessed, stored abroad or taken back entirely.

Third World countries, and particularly those just beginning to develop their nuclear industries, are by and large most interested in having their spent fuel safely managed or removed, although many would like compensation in cash or kind for the value of the plutonium (IAEA, 1978) The notable exceptions are Argentina and Brazil, which will probably be seeking domestic or bilateral solutions.

VI. Will national storage capacity be expanded?

The answer to this may be 'no' in a surprising number of cases. For FR Germany and the USA, internal debate may lead to expanded AR storage at first, but that is a minor issue. More important is the prospect that several states might choose not to expand domestic spent fuel storage capacity for fear that fuel suppliers would then impose restraints on fuel transfers or refuse to grant exemptions to existing restraints.

At present, Australia, Canada and the USA, the major suppliers of fuel outside the CMEA area, have placed restrictions on movements of the fuel they supply. The USA is at present the major enricher of LWR fuel and thus in a particularly powerful position; at the same time, it has the most restrictive set of conditions and provisions for retransfer under the terms of the Nuclear Non-Proliferation Act of 1978 (Donnelly, 1979). If countries such as Japan and Sweden did not have adequate spent fuel storage capacity, fuel suppliers would be faced with the choice between granting retransfers for reprocessing or forcing the nuclear industry of a valued ally to be shut down. South Korea and Taiwan, who gave up their own reprocessing plans at the insistence of the USA, are particularly well placed to attempt such leverage. The situation of the CMEA countries *vis-à-vis* the USSR appears to be structurally similar, but it is doubtful that they could credibly threaten either to reprocess domestically or to shut down if the USSR does not take spent fuel back. A CMEA central store might be feasible, but national storage seems most probable.

Of the other developed consumer countries, the ones most likely to ensure adequate storage capacity are those such as FR Germany and Switzerland who place a high value on not being overly dependent upon reprocessors. For the Third World countries, it is simply too soon to tell. They will be under little pressure to decide for several years to come.

VII. Will suppliers take spent fuel back?

The USSR has fuel-return agreements with its CMEA partners, although it is not sure how much will be taken back or when. As for Australia, Canada and South Africa, it seems very unlikely that they will take spent fuel back. Except for the USA, other suppliers of uranium are generally Third World countries without facilities to extend any offer.

The USA is in a key position, both because it is still the major enricher of fuel outside the CMEA area and because it has been the most concerned party. Indeed, the Nuclear Non-Proliferation Act of 1978 not only restricts transfers, but mandates a search for international spent fuel storage mechanisms (Donnelly, 1979). However, the USA will probably accept only small amounts of foreign spent fuel, even if Congressional approval is forthcoming. Since related proliferation worries have been centred on the Pacific Basin for the moment, an effort is being made by the USA to promote a Pacific storage facility (for Japan, South Korea, Taiwan and, eventually, the Phillipines) on some trust or territorial island. However, this has been opposed by Australia and New Zealand. An attempt to select the tiny island of Palmyra was a dismal political *faux pas*, and the search for a site continues. It will not be easy.

VIII. Future prospects

Storage of spent fuel appears to be a relatively straightforward technical problem, and international management, control or storage itself a comparatively simple institutional undertaking. It is the underlying politics of reprocessing that complicates matters internationally, just as public concern over all aspects of waste management complicates matters domestically. The most obvious solutions (listed below) seem at best improbable.

1. Storage in nuclear weapon states would 'solve' the proliferation problem: France and the UK would probably agree, for a fee, but only if tied to reprocessing contracts; China is out of the question; the USSR is unlikely to extend leasing outside the CMEA area, and the USA is unlikely to take back much.

2. An internationalized storage depot, or one under nuclear weapon state control, would be an attractive alternative. But there is some doubt that a site could be found that would be agreeable to all parties, if at all.

There is also some question as to whether the proliferation risk presented by spent fuel warrants such politically drastic solutions. There are a number of alternatives that are far less rigidly structured (Rochlin, 1979): (*a*) improved international safeguards, such as real-time or near real-time monitoring, (*b*) international, multilateral or bilateral supervision or control of national storage sites, and (*c*) regional siting under the aegis of an international spent fuel storage authority empowered only to set and enforce standards. Moreover, there is every reason to believe that the type of solution required will differ markedly from region to region. Both industrial and proliferation considerations are quite different in Europe from what they are in South America or the Pacific Basin.

The interweaving of industrial and energy policies, attitudes towards reprocessing and plutonium use, domestic antipathy to storing nuclear wastes from abroad, and non-proliferation policies and initiatives create a complex negotiating situation. Proposals that constrain individual behaviour too much are unlikely to be accepted. Proposals that do not constrain it enough may not satisfy the fuel suppliers who originated the idea. The outcome, if indeed there is one general enough to be called international, is more likely to lie towards the less-constraining end, with a possible major and expanded role for the IAEA. But it is difficult to visualize any general solution independent of a widespread agreement on the distribution and control of reprocessing and separated plutonium.

References

DOE (US Department of Energy), 1979a. *Draft Generic Environmental Impact Statement on Management of Commercially Generated Radioactive Wastes*, Report DOE/EIS-0046-D (NTIS, Springfield, Virginia).

DOE (US Department of Energy), 1979b. *Technology for Commercial Radioactive Waste Management*, Report DOE/ET-0028 (NTIS, Springfield, Virginia).

Donnelly, W.H., 1979. Applications of US non-proliferation legislation for technical aspects of the control of fissionable materials in non-military applications. In *Nuclear Energy and Nuclear Weapon Proliferation* (Taylor and Francis, London, 1979; Stockholm International Peace Research Institute), chapter 9, paper 14.
Hannerz, K. & Segerberg, F., 1979. Proliferation risks associated with different back-end fuel cycles for light water reactors. In *Nuclear Energy and Nuclear Weapon Proliferation* (Taylor and Francis, London, 1979, Stockholm International Peace Research Institute), chapter 3, paper 5.
IAEA (International Atomic Energy Agency), 1978. *International Management of Plutonium and Spent Fuel* prepared by the Secretariat with the assistance of expert consultants (IAEA, Vienna).
Johnson, A.B., Jr., 1977. *Behavior of Spent Nuclear Fuel in Water Pool Storage*, Battelle Pacific Northwest Laboratory Report BNWL-2256 (Battelle, Richland, Washington).
Marshall, W., 1978. Nuclear power and the proliferation issue, Graham Young Memorial Lecture, 24 February.
Miller, M., 1978. *International Management of Spent Fuel Storage: Technical Alternatives and Constraints*, MIT Energy Lab Report MIT-EL-78-012 (MIT Energy Laboratory, Cambridge, Massachusetts).
OECD (Organisation for Economic Co-operation and Development), 1977. *Reprocessing of Spent Nuclear Fuel in OECD Countries*, a report by an Expert Group of the OECD Nuclear Energy Agency (OECD, Paris).
Rochlin, G.I., 1979. *Plutonium, Power, and Politics: International Arrangements for the Disposition of Spent Nuclear Fuel* (University of California Press, Berkeley and Los Angeles).
Williams, F. & Deese, D., (eds.), 1979. *Non-Proliferation: The Spent Fuel Problem* (Pergamon, New York).
Windscale Inquiry, 1978. *The Windscale Inquiry: Report by the Hon. Mr. Justice Parker* (Her Majesty's Stationery Office, London)

Paper 18. Regional planning of the nuclear fuel cycle: the issues and prospects

B.W. LEE*

Atomic Energy Commission, Ministry of Science and Technology, Seoul 110, Korea

I. Introduction

In order to foster world peace, the non-proliferation objective must be achieved both horizontally and vertically. On the one hand the sensitive elements of the nuclear fuel cycle should be internationalized, and on the other hand the nuclear arms race should cease and effective international controls should be implemented for the speedy conversion of all military nuclear programmes to civilian programmes.

The present non-proliferation régime is effective in preventing neither the wider dissemination of nuclear weapons horizontally nor the rapid escalation of existing nuclear stockpiles vertically.

International measures to limit the proliferation of nuclear weapons rest upon a multitude of interactions among treaties and international statutes. Among these measures, the most far-reaching and comprehensive is the NPT. However, the NPT is discriminatory, and proliferation can best be prevented by political rather than by technical or institutional means (Goldschmidt, 1977). The aim of the NPT is the prevention of the emergence of a new nuclear weapon state. Therefore, it is essentially a treaty for the 'non-acquisition' of nuclear weapons rather than for non-proliferation. The Treaty does not prevent the nuclear weapon parties from increasing their nuclear arsenals nor from testing and improving weapon quality. For this reason, the NPT is said to relate to horizontal rather than to vertical non-proliferation (SIPRI, 1975).

In order to augment the present horizontal non-proliferation régime, multinational efforts are being made through the Export Guidelines of the London Club, the augmentation of IAEA (International Atomic Energy Agency) safeguards, the Vienna Convention on Physical Protection of

*The opinions expressed in this paper are those of the author and do not necessarily reflect the official policy of his professional affiliation.

Nuclear Materials, and the International Nuclear Fuel Cycle Evaluation (INFCE).

On the other hand, the unilateral efforts of the United States, based on the Nuclear Non-Proliferation Act of 1978, are being implemented through bilateral agreements for co-operation. The very foundation of such co-operation should be based on a spirit of mutual trust and confidence. Without such confidence, any international régime is doomed to failure.

Because of the multitude of complex problems involved in achieving the horizontal non-proliferation objectives, the concept of multinational participation in regional fuel cycle planning has been proposed on a number of occasions to augment the existing non-proliferation régime. By reviewing the development of this concept, the major issues may be identified and the desirability and practicability of proposed measures can be examined.

Since there are sufficient uranium reserves, enrichment capacity and fabrication facilities available up to the turn of the century, the front end of the fuel cycle is excluded from the scope of this paper. Reprocessing is discussed as an example of the most controversial issues. If the most difficult part of the fuel cycle could be resolved by multinational participation as a form of internationalization, the remaining fuel cycle issues could be easily undertaken.

II. The desirability of regional planning

From the late 1960s, and particularly after the oil embargo in 1973, the Republic of Korea sought an alternative energy strategy due to a deficiency in traditional energy resources. In 1974 the *Long-Range Nuclear Power Program Study* identified nuclear energy as the most practicable alternative energy resource to meet the Republic of Korea's rapidly growing demand. Nuclear energy can help to bridge the energy-supply gap until a stable and abundant new source becomes commercially available some time in the next century.

In the long-term planning of any nuclear power programme, an assured nuclear fuel supply and assured availability of cycle services are of the utmost importance. For actively developing countries, such as the Republic of Korea, without indigenous energy resources, the growth rate of nuclear power is very steep, and the economies of scale may well justify national reprocessing centres becoming more appropriate, due to the large size of nuclear power programmes. In their initial stages, however, regional reprocessing centres will serve their objectives until national centres become practical. As a consequence, bilateral and trilateral discussions regarding regional reprocessing centres were initiated in the Far East with a view to future multinational participation. A conclusion of the *Long-Range*

Nuclear Power Program Study was as follows: "A spent fuel reprocessing plant in Korea would be feasible only through the participation of other Asian countries in a regional program in which a reprocessing plant would be located in Korea to serve nuclear power reactors in Korea, Japan, Taiwan, the Phillipines, and other Asian countries" (Kaiser Engineers, 1974). Based on this conclusion, a call for the regional planning of spent fuel reprocessing and waste management was first made at the Eighteenth General Conference of the IAEA by the delegate of the Republic of Korea, as follows:

When the role of nuclear power becomes more important, the world will be faced with the grave problem of what to do with spent fuels and radioactive wastes. In the future, the transporting of spent fuels long distances across the world to reprocess them will be not only uneconomical but almost impossible. This would be a good area for possible regional co-operation. Nearby countries could get together with the common goals of spent fuel reprocessing and waste management. I would hope that the IAEA will sponsor such a regional co-operation project, which would benefit most developing countries. (Choi, 1974)

This call was well received, and the preliminary study on the IAEA Regional Fuel Cycle Centre (RFCC) Study Project was undertaken, as the Director General explained to the UN General Assembly in November 1974:

Economic grounds alone require that fuel reprocessing be done on an international basis, perhaps by regional plants. Since reprocessing plants are the source of 99 per cent of nuclear wastes, there are also obvious advantages from a safety point of view in minimizing the number of reprocessing plants and waste storage sites. Prevention of diversion also becomes easier if reprocessing plants are few in number and operated under regional or international auspices. This consideration and the problems of physical security would also encourage locating fuel fabrication plants. (IAEA, 1977)

The IAEA RFCC study, started in 1975 and concluded in 1977, was comprehensive, both in participation and in coverage. Consultants from member states and relevant international organizations dealt with irradiated fuel from the time of its discharge from power reactors through storage, reprocessing and the fabrication of new mixed-oxide fuel elements; the study also dealt with radioactive waste management.

At the first NPT Review Conference in May 1975, the desirability of regional planning for key nuclear fuel cycle facilities was reaffirmed by the Korean delegate, as follows:

Because of the capital as well as technical intensiveness of the key nuclear fuel cycle facilities, most developing countries cannot afford to build commercially competitive facilities by themselves, simply because of the smallness of the justifiable nuclear power program, compared to advanced countries. Unfortunately, the major commercial key fuel cycle facilities are located outside of these areas....this is the area in which my delegation would like to endorse the necessity for a regional and multinational fuel cycle center...(Lee, 1975)

The Declaration of the first NPT Review Conference recognized "that regional or multinational nuclear fuel cycle centres may be an advantageous way to satisfy, safely and economically, the needs of many states in the course of initiating or expanding nuclear power programmes, while at the same time facilitating physical protection and the application of IAEA safeguards, and contributing to the goals of the Treaty". (IAEA, 1977).

At the Nineteenth Regular Session of the IAEA General Conference in September 1975, the Director General stressed the long-range planning and international co-operation potential of the RFCC Study Project. A number of IAEA member states, among them the delegation from the Republic of Korea, responded with statements of full support.

III. Practicability—issues and prospects

The success of multinational participation in regional planning really depends on how major issues are resolved. At the first Pacific Basin Conference on nuclear power development and the fuel cycle in October 1976, a number of practicable approaches for regional planning were presented by participants from the Far East and from IAEA RFCC Study Project teams.

At the outset of the RFCC study, the USA was a strong supporter of multinational control of fuel reprocessing due to its non-proliferation merits. The conclusion of the study—that economies of scale might provide sufficient incentive for states to join together without political pressure—was, however, not received with much enthusiasm by the USA. During the intervening time, concern that participation in an RFCC might accelerate the diffusion of technical capabilities, thus aiding in the construction of national plants, resulted in the withdrawal of US support (Rochlin, 1979). The IAEA was cautiously critical of such a policy development, as reflected in the statement to the Twenty-first Regular Session of the General Conference of the IAEA by the Director General:

> ...There is recognition of certain limitations of the non-proliferation measures so far taken....Even the strictest international safeguards verification does not prevent accumulation of weapons-grade nuclear material within the peaceful fuel cycle. Hence there is a tendency to try to prevent proliferation by limiting the expansion of the peaceful nuclear fuel cycle. There is also an awareness that such measures might be counter-productive by impeding development or by encouraging independent national fuel cycles instead of furthering international co-operation in this field.
>
> It is worth recalling that reprocessing was declassified by the time of the first Geneva Conference in 1955. As the result of experience gained to date, including the Agency's study of regional nuclear fuel cycle centres, it is generally accepted that the number of such plants should be limited to a minimum. To prohibit them, however, would probably lead to a result opposite to that intended. Isotope separation has always been classified by the nuclear weapon States, which only seems to have

stimulated a great deal of work in several countries on new separation methods, a fact which became clear at Salzburg. This is a prime example of how a policy of denial may stimulate research and development activity in a sensitive area. Let us remember that in the long run there is no way of stopping the spread of nuclear technology amongst nations, and we must face the proliferation problems that result. The question is therefore not how to stop nuclear development but how best to make use of it and how to apply effective safeguards. (Eklund, 1977).

In any new practical international régime there are bound to be both merits and shortcomings. The inevitable diffusion of sensitive technical capabilities through multinational participation in an RFCC is a necessary evil. However, regionalization of fuel cycle centres offers several advantages. It can bring economy of scale and reduce potential environmental release points. With regionalization, a greater concentration of funds and expertise is available to devote to such matters as process improvement, health and safety, and environmental protection.

One of the most important benefits is the net positive effect on non-proliferation. In the absence of regional facilities, each country is forced to develop its own fuel re-processing capability, which can only foster proliferation. Unilateral denial of sensitive technologies can only delay this process temporarily, but in the long term would stimulate research and development efforts and diffusion of technical capabilities.

The balancing of these factors and a final evaluation should be made in the near future. However, the advantages in regard to non-proliferation considerations, health, safety and environment as well as the technical and economic aspects may well outweigh the risk seen in the inevitable diffusion of some technological capability. Moreover, a positive approach would enhance the atmosphere of mutual trust and confidence among the participants. A policy of denial can only create distrust and mutual suspicion and further jeopardize the peaceful uses of nuclear energy.

For multinational participation in regional fuel cycle planning, a number of issues must be resolved, such as security of supply, access to services and technology, international safeguards, physical security, health and safety, environmental protection, location, site selection, participation, operation, ownership, management, control of facilities, financing, host-country responsibilities and rewards, insurance and liability, and socio-political issues among potential partners.

In view of the different attitudes expressed in INFCE and while awaiting the results of the second NPT Review Conference, it would be too early to offer further judgement on the practicability of multinational participation in regional fuel cycle planning, particularly for reprocessing. It appears that the success of the concept will be affected by INFCE, the NPT Review Conference and possibly the proposed UN conference on nuclear energy. Above all, the potential influences of these international forums will demonstrate the need to recognize that, in such a new international nuclear order, mutual trust and confidence will be the essential ingredients.

References

Choi, H.S., 1974. Statement to the Eighteenth Regular Session of the General Conference of the IAEA by the Korean Delegate, in Vienna, September.

Eklund, S., 1977. Statement to the Twenty-first Regular Session of the General Conference of the IAEA by the Director General, in Vienna, 26 September.

Goldschmidt, B., 1977. Goldschmidt calls fuel cycle access best way to limit proliferation, *Nucleonics Week* 18 (14): 8–9.

IAEA (International Atomic Energy Agency), 1977. *Regional Nuclear Fuel Cycle Centers,* vols. 1–3 (IAEA, Vienna).

Kaiser Engineers and Constructors, Inc., 1974. *Long-Range Nuclear Power Program Study,* Executive Summary, Report 74-101R, 6.

Lee, B.W., 1975. Statement to Committee II, First Review Conference on NPT by the Korean Delegate, in Geneva, 20 May.

Rochlin, G. I., 1979. *Plutonium, Power and Politics—International Arrangements for the Disposition of Spent Nuclear Fuel* University of California Press, Berkeley), p.165.

SIPRI (Stockholm International Peace Research Institute), 1975. *Preventing Nuclear-Weapon Proliferation: An Approach to the Non-Proliferation Treaty Review Conference.* SIPRI brochure.

Paper 19. Multinational arrangements for enrichment and reprocessing

I. SMART

Ian Smart Limited, 3 Grosvenor Avenue, Richmond, Surrey TW10 6PD, UK

I. Introduction

In considering multinational arrangements for the nuclear fuel cycle, a fundamental distinction must be drawn between arrangements to control materials and arrangements to exploit industrial processes. To those concerned with proliferation, controlling sensitive materials may seem to offer a quicker return. In the long run, multinational process arrangements may have a no less important role. They will also, however, be more difficult to establish.

The proliferation of nuclear weapons is a result of producing nuclear explosives. Such explosives require fissionable materials—plutonium or high-enriched uranium—which may be present in a civil nuclear fuel cycle. But explosives themselves are not produced by any civil fuel cycle process. Fuel cycle processes are 'sensitive', therefore, only to the extent that they may produce sensitive (that is, explosive-usable) material. Their relative sensitivity tends, moreover, to be an attribute of specific industrial plants, rather than that of the processes themselves, and to depend on particular circumstances, especially of location. In contrast, fissionable materials, readily transportable and exploitable, tend to be sensitive *per se*, regardless of circumstantial factors. One implication is that, while international efforts to control sensitive materials can be aimed at creating a notionally general régime, proposals for multinational process arrangements must be related, with inevitable difficulty, to particular plants in particular countries.

II. Economic and industrial pressures on process arrangements

Whereas the cost of secure facilities to store material is commonly small in relation to total costs of nuclear power, enrichment and reprocessing plants entail very large capital and operating costs. They also involve the efficient application of complex technologies, mastery of which represents a major economic asset to those who have attained it.

For these reasons, the pressure to obtain an economic rate of return on multinational enrichment or reprocessing activity is much higher than in the case of material control. If alternative multinational material and process arrangements were seen to have a similar effect on the probability of nuclear proliferation, then the perceived cost of obtaining that effect through process multinationalization, involving heavier capital and revenue liabilities and some dissipation of technological assets, would be greater. That greater cost could be offset only by the economic return on enrichment or reprocessing itself.

Economic considerations are all the more powerful because multinational process arrangements demand the involvement of industrial, as well as governmental, agencies. In the case of material control, there is no reason why governments wishing to create a multinational régime should not operate it themselves, as a joint administrative rather than industrial activity. Even if enrichment and reprocessing were a state monopoly everywhere—which is not likely to be the case—it is questionable whether multinational arrangements for the industrial conversion of materials could serve administrative criteria alone. As it is, they must not only be measured by industrial standards, but also be supported by groups from both private and public sectors. Again, this means that they must be seen to serve an economic, as well as a political, purpose.

Governments have good reason to incur costs in return solely for reducing the probability of proliferation. However, industries, and especially private industries, cannot be expected to accept the costs of their necessary role in multinational process arrangements on that basis alone, given their additional responsibilities to shareholders and customers. Their net economic return, however broadly conceived, must also be at least unimpaired, and if possible increased.

III. Non-proliferation benefits

Obviously, multinational fuel cycle arrangements cannot prevent proliferation, any more than international safeguards can do so, if only because a country determined to produce nuclear explosives, or to keep a weapon option open, will either stand aside or withdraw from them. However, as a complement to safeguards, multinationalization can present an additional deterrent to proliferation, in that a country subscribing to it must expect to incur some political penalty if it violates or renounces multinational obligations in order to obtain nuclear explosives. In that respect, multinational process arrangements may be more effective than those dealing only with the control of materials, since the abrogation of a process arrangement by any one partner is likely to impose real and visible costs on others, whereas unilateral abrogation of a material control arrangement, except perhaps by a host country, is unlikely to damage the interests of others to a comparable extent.

Host countries occupy, of course, a special position, but it would be wrong to see it as a position of unmitigated power. A country acting as host to a multinational enrichment or reprocessing facility might have most to gain from abrogating the agreement concerned, since it would then have control of the facility. But it would also have most to lose, in that its action would most significantly damage the interests, and thus attract the wrath, of its partners. The probable effect of expropriating from other national governments or enterprises might therefore deter a host state more strongly than the probable effect of expelling or excluding safeguards inspectors.

A country which subscribes to a multinational process arrangement associated with adequate non-proliferation controls is seen to be more strongly deterred from violating those controls. One corollary is that such an arrangement offers a government which wants to demonstrate its non-proliferation commitment to suspicious outsiders an additional and persuasive means of doing so—and, by doing so, of diluting any fear that it may seek nuclear explosives in the future.

Because it is the fear of future proliferation, at least as much as the event itself, which strains international stability, containing that fear has been a vital objective of non-proliferation policies. In many cases, a government could dispel international fear of its future intentions by accepting safeguards, especially in the context of the Non-Proliferation Treaty (NPT). If a residue of fear persisted, it might then perhaps be dissipated by an additional commitment to deposit sensitive materials in international custody. However, in some cases, as we know, even that would not be enough.

Independent enrichment or reprocessing, albeit under safeguards, by a country whose behaviour, predicament or ambition has aroused the apprehension of others may still generate fear that civil nuclear technology will in future be converted to military purposes. Subscription to a multinational arrangement which includes a credible political deterrent to independent activity may be the best way available of demonstrating that such fear is unfounded. The requirement that the arrangement must also be economically useful is likely still to apply. In such a case, however, the admitted difficulty of constructing a multinational process arrangement may readily be justified.

IV. Economic and industrial benefits

The fact that multinational process arrangements can be economically advantageous is shown by actual cases of multinational enrichment or reprocessing: Eurochemic or United Reprocessors, and Urenco or Eurodif. Each represents a particular model of multinationalization, just as each represents a response to particular needs. Eurochemic was conceived primarily as a mechanism for international communication on reprocessing technology, whereas United Reprocessors has played its principal role in relation to the commercial market. Urenco was formed to multiply the technical and commercial strength of three notionally equal partners, while Eurodif serves rather as a means of mobilizing multinational investment in the exploitation of enrichment technology by one state. All have nevertheless been judged to benefit the economic and industrial interests of multinational partners. Some may have an incidental non-proliferation effect as well: for example, Urenco, through the provision in the Treaty of Almelo which precludes separate commercial use of centrifuge enrichment technology. Each may also have features potentially applicable to multinational arrangements designed to counter the fear of proliferation. Primarily, however, they stand as proof that, in appropriate circumstances, multinational process arrangements can offer economic and industrial advantages over purely national activity.

V. Criteria for fuel cycle multinationalization

Restraint

In the present context, proposals for fuel cycle multinationalization must impose some credible limit on the perceived probability—and thus the fear—that any participating non-nuclear weapon state will convert a civil nuclear programme to military ends. As that implies, any such proposal is to be judged, in the first instance, by a criterion of restraint.

Subscription to a multinational process arrangement must entail a voluntary but substantial restriction of the right to enrich or reprocess nationally and must include formal commitments to that effect. It must also provide participating states with credible means, opportunity and motive to verify each other's observance of those commitments. Multinational involvement in directing and monitoring industrial activity under the arrangement must therefore be sufficient to persuade both members and non-members that clandestine violation of non-proliferation commitments is implausible. At the same time, the membership itself must be so composed as to rule out any reasonable suspicion of multinational collusion or connivance in such a violation.

Merely symbolic multinationalization is thus to be ignored, as is any arrangement limited to multinational investment in enrichment or reprocessing by one country, or to multinational co-operation in marketing such a service. To satisfy the criterion of restraint, a multinational process arrangement must constrain industrial activity itself. How far multinational staffing need go for that purpose is a matter for detailed negotiation. Clearly, however, the employment of process plant and the disposition of its product must be subject to overall multinational determination, and there must be a sufficiently unrestricted degree of multinational access to the plant itself.

Viability

While the need for proliferation restraint is self-evident, the contention of this paper is that multinational process arrangements, to be acceptable in practice, must also make good economic and technical sense: that they must, in fact, satisfy a criterion of viability. The practical benefits offered to potential participants, when all direct and indirect costs have been counted, must therefore be at least equal to those available through national enrichment or reprocessing programmes.

It follows that a multinational fuel cycle enterprise must show itself capable of competing commercially in the relevant market. This has numerous implications. Clearly, for instance, a multinational enterprise cannot be so limited in its choice of technologies as to cripple its commercial effectiveness, especially when access to a more efficient enrichment or reprocessing technology may constitute one of the important incentives to join the enterprise in the first place. Nor can the physical location of plants be restricted in a way which imposes penal costs, for transport or otherwise. Nor, indeed, can multinational supervision of industrial operations, however essential, place an economically disabling burden on productivity, quality, safety or price.

The hope must rather be that process multinationalization, far from imposing economic penalties, will offer economic advantages. A multinational process enterprise, serving several national nuclear power programmes, should be able to sustain a larger enrichment or reprocessing capacity than any single participant alone. The principal advantage is therefore that of lower unit cost, achieved through scale economies. The extent to which enrichment or reprocessing plants are responsive to conventional scaling rule, relating unit capital cost to plant size, is a debatable matter. What is certain is that the higher plant utilization which a well-designed multinational process arrangement should ensure, and the lower unit costs which that would entail, represent a potentially important benefit.

Symmetry and parsimony

In designing multinational process arrangements, the criteria of proliferation restraint and industrial viability are to be considered co-equal. Unless the non-proliferation and industrial benefits offered are broadly commensurate, those pre-eminently concerned with one aspect or the other will suspect an attempt to exploit them. The necessary relationship between restraint and viability implies, therefore, a subsidiary criterion of symmetry. With that in mind, it will sometimes be essential to import elements into the arrangement which are not strictly relevant to its original purpose but which have to be included to provide a balanced set of benefits. That necessity must be approached, however, with caution. Given the difficulty of establishing and sustaining multinational arrangements, involving separate partners with independent interests, their operational scope must be as narrow as the criteria of restraint, viability and symmetry will permit. There is, in fact, a second subsidiary criterion of parsimony to be respected.

The practical problem of satisfying criteria of symmetry and parsimony should not be underestimated. The best, in terms of either non-proliferation or industrial economics considered separately, may be the enemy of the good, in terms of an acceptable and practicable arrangement. To take reprocessing, there is, for example, an industrial argument for extending a multinational arrangement to cover final disposal of radioactive waste generated as a by-product. Conversely, there is a non-proliferation argument for extending the arrangement to cover recombination of separated plutonium into either a standard mixed-oxide (MOX) material or fabricated MOX-fuel elements before release into national custody. In each case, however, committing the arrangement to such an extension may be rash.

Although the issue is irrelevant to non-proliferation, the opportunity to transfer responsibility for high-level radioactive waste to a multinational entity would be an additional incentive to many countries to join a multinational reprocessing arrangement. Unfortunately, it might also make it impossible to find suitable sites for multinational facilities. As to the re-combination of separated plutonium, the non-proliferation reasons for seeking it may seem obvious. In practice, however, the cost and difficulty of fabricating reactor-specific MOX fuels for a wide international market are likely to be excessive, while the problem of producing a basic MOX material suitable for reblending to varying uranium/plutonium ratios, without prior reseparation of plutonium, turns out to be peculiarly obstinate. At the same time, although plutonium contained in MOX may be less vulnerable to theft by private groups, the wider non-proliferation benefit of releasing it into national custody only in that form turns out to be small. There may be cases, therefore, in which the criterion of parsimony is of special significance.

VI. Technology transfer

The most difficult problem in satisfying the criteria applicable to multinational process arrangements may well be that of technology transfer. The basic design of a plant for reprocessing by the Purex method is widely known. The difficult industrial techniques needed to operate a commercial reprocessing plant safely and efficiently are not. Conversely, whereas the operation of a diffusion or centrifuge enrichment plant presents relatively familiar problems, crucial design features are, in both cases, known to relatively few countries. It follows that preventing the dissemination of either reprocessing or enrichment technology has been widely seen as an objective of non-proliferation policy. However, it also follows that, for many countries, access to that knowledge represents a potentially important industrial incentive to join a multinational enterprise. Moreover, a sufficient understanding by all partners of the technology involved is likely to be needed if the arrangement is to convey an assurance that non-proliferation restraints are being collectively enforced. Yet some of the technology involved has a proprietary value and is subject to rules of commercial confidentiality. The problem of determining whether or when ostensibly sensitive technology is to be shared among multinational partners when the criteria of restraint and viability seem thus to be in conflict is therefore peculiarly difficult.

No general solution to the problem of technology transfer under multinational process arrangements is plausible. The issue will have to be negotiated on a case-by-case basis. Some general considerations are nevertheless relevant. One, for example, is that agreement on handling technology transfers is likely to be easier if the defined scope of sensitive technology is kept to a minimum. Since any country participating in a multinational process arrangement designed to reinforce a non-proliferation régime must be presumed to have all its civil nuclear activities under international safeguards, there is no purpose in attempting to impose additional non-proliferation restraints on any but the uniquely sensitive technologies of enrichment and reprocessing: that is, on technology capable of being the proximate source of explosive-usable material.

A second consideration is that Article IV of the NPT has already imposed a general obligation on parties "to facilitate...the fullest possible exchange of equipment, materials and scientific and technological information for the peaceful uses of nuclear energy". Especially in the case of NPT parties, therefore, it is on the shoulders of those who possess relevant technology that there lies the onus of proving that it should not be shared within a multinational process arrangement. Politics makes that burden heavier. It is universally apparent that access to industrial technology represents a mark of international status, and that denial of access provokes resentment and friction between more and less industrialized states. Multinational fuel cycle arrangements are unlikely to be successful unless proposals for them escape such international conflict over development and status.

The vexed issue of technology transfer has an air of unreality. On the one hand, many arguments for withholding sensitive fuel cycle technology seem to assume that a technical oligopoly can be sustained indefinitely. On the other hand, experience points to the fact that there is ultimately no limit to the number of states capable of mastering nuclear technologies, including those of enrichment and reprocessing. The real issue, therefore, is one of time.

An unqualified policy of withholding sensitive technology may delay its wider dissemination, but it is also likely to make multinational fuel cycle arrangements more difficult, or even impossible. The critical choice is thus not between the dissemination and non-dissemination of sensitive technology. It is whether, from the non-proliferation point of view, the gradual but relatively uncontrolled diffusion of enrichment and reprocessing technology, leading to the establishment of more independent national plants, is to be preferred to the deliberate, possibly swifter, but certainly more regulated transfer of technology under multinational auspices, in return for the assurance that its exploitation will also be under multinational control.

VII. Practical implications of the market

However well proposals for multinational process arrangements meet general criteria or reflect general considerations, their chances of acceptance will depend critically on how well they match the circumstances of civil nuclear activity in the real world. In the case of enrichment, for example, an examination of prospective supply and demand shows that plants in being or firmly planned (in Brazil, France, FR Germany, Japan, the Netherlands, South Africa, the Soviet Union, the United Kingdom and the United States) will have enough capacity to satisfy world needs until 1990, and probably until 2000. There seems to be little prospect, therefore, that any proposal for multinational enrichment based on constructing additional capacity would be considered viable within the next 10–20 years. If a multinational enrichment enterprise is to be launched successfully, it will have to be built on plants already in existence or planned.

In contrast, the world reprocessing capacity may be chronically deficient between now and the end of the century, at least in relation to the rate of spent fuel arising. How significant that will be will obviously depend on the extent to which an energy demand for plutonium emerges. It is not impossible, however, that a viable multinational reprocessing enterprise could be founded upon either existing and planned plants or the construction of additional capacity, or upon a combination of both.

VIII. Institutional forms

As with those multinational enrichment and reprocessing consortia already formed for industrial reasons, future multinational process arrangements made with non-proliferation in mind are likely to vary widely in institutional form. In all probability, however, the dual criteria of restraint and viability will require a demarcation between political supervision and the direction of industrial policy. Characteristically, the former role might be played by a council of political representatives from the states concerned, while the latter would be the responsibility of a supervisory board in which industrial representation would prevail. There might also be a central marketing body. All other functions, including the immediate management of enrichment or reprocessing plants, would be delegated to operational subsidiaries, each embodying a balance of investment and policy influence appropriate to the individual plant concerned. The overall purpose would be to ensure the equal application of basic standards, in regard to both non-proliferation and industrial policy, while offering participating countries and industries the sort of autonomous choices about investment, technology and industrial management which would convince them that multinationalization did not diminish their economic security.

IX. Flexibility and security

In one form or another, security is, in fact, the key to the success of any multinational fuel cycle arrangement. Those concerned to reduce the probability of proliferation will look to such arrangements to reinforce international security. But those concerned to develop peaceful uses of nuclear energy will also demand that any multinational agreement should serve security. In some cases, it will be security of supply or security of deposit, both of which are plausible functions of multinational arrangements for material control. In the case of multinational process arrangements, however, the security sought by participants will be security for their industrial programmes and energy plans. The danger they must guard against is that of exploitation or economic disablement, and the security they seek must come from confidence that they can choose the 'best' product or process, at the 'best' price, in the context of a market.

There is no doubt that well-designed multinational process arrangements can reinforce a non-proliferation régime. The test of their acceptability and usefulness is likely to be rather whether they can also be flexible enough to offer a demonstrably efficient degree of technical and commercial choice. If they cannot do so, potential participants in multinational arrangements will instead create that sort of choice by embarking upon enrichment or reprocessing on a national basis.

Paper 20. Sanctions as an aspect of international nuclear fuel cycles

P. SZASZ*

Office of the Legal Counsel, Office of Legal Affairs, Room 3440, United Nations, New York, New York 10017, USA

I. Introduction

Every abstract study of the requirements of an international nuclear control system has recognized that one important aspect of such a system would be the imposition of appropriate sanctions in the event of a wilful violation of the obligations to which the system relates. Nevertheless, none of the several control systems actually in operation—those of the International Atomic Energy Agency (IAEA), Euratom or the Nuclear Energy Agency—appears to have any effective sanctions, a point that has not escaped those critics that consider these systems to be largely ineffective.

One response to such a criticism might be the assertion that the importance of sanctions must not be overemphasized as if they constituted an absolutely essential ingredient of any control system. Indeed, it can be argued that the mere installation of such a system may suffice in respect of those states that voluntarily participate in it, for governments will generally not enter into or remain bound by any agreement with which they do not wish, or may not be able, to comply (Chayes, 1972). More important in respect of ostensibly peaceful nuclear activities, control measures (safeguards) that are considered to be certain to detect a violation will more than likely deter such violation, as governments in any event abhor being caught violating their international obligations.[1] Finally, the detection alone of any undeterred violation may suffice to protect the international community, by enabling states, individually or collectively, to take measures to protect their interests, even without imposing formal collective sanctions designed to ensure compliance. Thus, even a sanctionless system is valuable

* The opinions expressed in this paper are those of the author and do not necessarily reflect the official policy of his professional affiliation.

[1] This is the principle on which the IAEA safeguards in implementation of the Non-Proliferation Treaty (NPT) are based. See IAEA document INFCIRC/153(Corrected), para. 28.

in relieving the world of the fears of secret or shadow (i.e., suspected but unconfirmed) proliferation.

However, it still remains true that a control system devoid of sanctions is perceived to be incomplete and thus ineffective. And whether or not such a perception is fully justified, peoples and governments will be unwilling to entrust their security in this crucial field to a system that appears to be inadequate. It is thus necessary to examine the reasons why effective sanctions have not been introduced into any existing nuclear system, what the specific obstacles to their introduction are, and whether these might not be overcome or mitigated by internationalizing part of the nuclear fuel cycle.

II. A control system with sanctions

To carry out the first part of the analysis proposed above, it may be useful to establish what the distinct, though interlocking, elements of any control system must be in order to permit the effective application of sanctions, and what might be the special obstacles to deploying these elements of a nuclear control system.[2]

Obligations

In order to contemplate penalizing a state for some particular conduct, it is evidently necessary that that conduct violate an important obligation of that state. In general, for such an obligation to exist, it must be one freely assumed by the state itself, though some obligations may be considered as deriving from overriding principles of international law from which no state can withdraw, and certain obligations might be created by a duly empowered international organ.

In respect of the peaceful uses of nuclear energy, many states have indeed already entered voluntarily into important obligations, as part of bilateral, regional or world-wide arrangements. In particular, many states have agreed either not to manufacture nuclear weapons and nuclear explosives at all or not to do so by the use of certain specially dedicated materials or facilities; in addition, in support of those primary obligations, these states have also obligated themselves to accept more or less strict control measures (for the most part safeguards administered by the IAEA). However, other states have failed to do so, and therefore it is important to note that, under current principles of international law, their failure to do

[2]This analysis constitutes a condensation of Szasz (1979). The sanctions discussed in this paper are those characterized as "formal collective sanctions" in the above article.

so does not violate any recognized overriding principles, nor does it seem likely that any such principles will gain sufficient general acceptance within the foreseeable future. On the other hand, there would seem to be no technical obstacle to the UN Security Council's declaring that nuclear proliferation—and even the conduct of uncontrolled peaceful nuclear activities liable to lead to such proliferation—constitute "threats to the peace" and are therefore proscribed (Szegilongi, 1978). So far, however, the Security Council has not done so, nor have any serious proposals been advanced to that end.

Safeguards

Once obligations are firmly and clearly established, the next link in the chain of control is to observe compliance with these obligations, that is, to implement 'safeguards', a term and concept well established in the nuclear field. During the past decades considerable experience has been gathered in the implementation of safeguards. It has been found that some facilities (e.g., certain types of reactor) are relatively easy to safeguard; that is, relatively inexpensive and non-obtrusive measures will suffice to establish with a high degree of certainty that no diversion is taking place. On the other hand, certain types of facilities, especially those that produce fissionable material in relatively pure form (e.g., enrichment and reprocessing plants) or that use it (e.g., fuel element fabrication), may require very intensive, expensive and obtrusive safeguards to preclude diversion; indeed, it is by no means clear yet whether the IAEA will actually be in a position to impose adequate safeguards in respect of large facilities of these types. Fortunately, however, it may not be necessary to observe a substantive misdeed directly, for if the safeguards are good enough to be able to detect it, the offending state is likely, well before such detection could take place, to give direct or indirect notice of its unlawful intent by preventing the further carrying out of controls.

Decision-making

Once the safeguards system is in place, it will from time to time receive indications that a controlled state may be violating or might have violated its obligations. Those indications must naturally be evaluated, in the first instance by the technical experts of the safeguards organ, but ultimately by an appropriate political (i.e., representative) organ. The task of that organ may not be an easy one, for the safeguards organ will most likely not be in a position to report an unambiguous detection of a diversion or an outright refusal to co-operate with the control procedures; rather there may be suspicious discrepancies in reports or excuses as to why inspectors are 'temporarily' being denied access to certain facilities. These ambiguities are likely to supply ammunition to those members of the decision-making body that are more or less closely allied or politically sympathetic to the accused state. Thus it may be difficult to secure—in particular, rapidly—a formal determination of a violation.

Once such a determination has been made, the same or some other representative body must decide on what sanctions are to be implemented by the other states participating in the control system and by the international organizations they control. These decisions too are fraught with technical and legal difficulties. In the first place, it may (as described below) be difficult to find any effective measures that can be practically implemented and would be likely to cause the offending state to change its conduct. Furthermore, states have historically been unwilling to exert serious pressures on other states in the collective interest, in particular if to do so is in any way burdensome for the imposing state. Especially allies of the offender, or those in fear of its retaliation, are likely to be most reluctant to vote for effective sanctions.

Imposition of sanctions

Difficult as some of the other stages of the control process may be, the greatest difficulty is likely to arise in the actual imposition of effective sanctions. Short of deploying overwhelming military force (a possibility that should not be excluded, but would have to be mandated by the Security Council pursuant to Article 42 of the UN Charter), experience shows that other merely political or economic sanctions are likely to be ineffective, at least in the short run, against a determined violator of an international obligation. For various legal, practical and political reasons, the sanctions measures most likely to be applied to a violation of a nuclear non-proliferation agreement are ones directed against the nuclear activities of the offending state. Aside from the peculiar appropriateness of doing so, for many states these activities may be the sector of their economy or technology most vulnerable to the deprivation of foreign inputs, and, if the country has become substantially dependent on nuclear power, pressures in that area may exert considerable leverage on the economy as a whole. However, a state may—and a state planning to defy an international control system most likely will—attempt to cushion the shock of such anticipated sanctions by stockpiling and by otherwise making itself as nuclearly autarkic as feasible. As a consequence, the sanctions most likely to be applied may prove, even if in the long run not completely useless, at least ineffective within the short time-frame considered relevant in respect of violations of a non-proliferation obligation. Although there may be other methods to prevent states from employing such sanctions-mitigating devices (e.g., collective supplier agreements to restrict nuclear shipments so as to prevent undue stockpiling), the most obvious way to do so would appear to be internationalization of part or all of the nuclear fuel cycle (see section III below).

Burden-sharing

The imposition of most types of sanction—with the exception of a mere cut-off in gratuitous assistance—is likely to be expensive, troublesome and possibly dangerous, to some or all of the states participating in the collective enforcement.

In respect of nuclear sanctions, these costs may be due to loss of business opportunities, the costs of transporting and storing any materials withheld or withdrawn from the sanctioned state, and the effects of any retaliatory measures (e.g., the failure to continue payments on materials and facilities previously supplied). In the normal course of events it is quite likely that these burdens will be capriciously distributed, and thus their main impact will not be borne either by those best able to do so or by those most interested in the success of the sanctions effort.

Consequently, a complete control system will also have to make provision for the sharing of any extraordinary costs in imposing sanctions, in addition to the normal costs of safeguards. Again, experience shows that the development of burden-sharing formulae is awkward and controversial and may even cause a breakdown of, or failure to achieve, an effective collective control system. Naturally, this danger is greater, the heavier and more uncertain the sanctions costs are likely to be. Consequently any method of reducing or mitigating such costs would enhance the possibility of plausibly threatening a state and of successfully imposing sanctions.

III. Sanctions and a partially internationalized fuel cycle

The final aspect of the analysis called for in this paper is to determine to what extent the imposition of sanctions might be facilitated by the internationalization of significant parts of the nuclear fuel cycle. (The different elements of a control system for such a venture are presented in the same order as in section II above.)

For the purpose of this analysis, it is assumed that only part of the fuel cycle will be internationalized—that is, conducted as a multinational operation by a multinational staff—and that part, presumably in any event the operation of power reactors, will remain national. Should the entire fuel cycle be internationalized, as foreseen in the original Baruch Plan, then the occasion for applying safeguards and sanctions, at least as now conceived, would fall away. Instead, certain other types of controls and precautions would have to be instituted in respect of any internationalized facilities. In particular, it would be necessary to ensure that: (*a*) a state ostensibly fully dependent on the international facilities does not clandestinely establish others for military purposes; (*b*) neither the host state nor any other seizes an international facility, and (*c*) the operations of international facilities are such that no fissionable material can be diverted to any state or other entity. All these problems, while important, are beyond the scope of the present paper. Nevertheless, it should obviously be noted that to the extent that more and more of the nuclear fuel cycle is internationalized, the scope for the application of conventional sanctions and other control measures is in any event reduced. The sections below indicate that for qualitative reasons the extent of this reduction is likely to be more significant than would be suggested by any mere quantitative weighing of the national and international portions of the cycle.

Obligations

It is necessary to base a control system on internationally valid obligations of the states concerned, whether the controls apply to the national part of a nuclear fuel cycle or to a fully national cycle. In the former case, it may be useful to add undertakings that states are to refrain from duplicating, on a national basis, those international facilities in which they are participating, for only to the extent that such undertakings can be secured will it be possible to attain certain benefits from the international system discussed below.

Safeguards

As indicated above, not all facilities are equally easy to safeguard. As it happens, those that may be most difficult to control, such as fuel-element fabrication and reprocessing facilities, are also the ones that are, for other reasons (principally economic), prime candidates for internationalization. If these facilities are internationalized and safeguards can concentrate on national nuclear reactors, this element of the control system will become much simpler and cheaper to implement, as well as more likely to be unexceptionably executed.

Decision-making

The principal task and structures of the decision-making mechanisms required for the imposition of sanctions do not require change because of the internationalization of part of the fuel cycle. Indeed, because such internationalization may, as discussed below, enhance the possibility of imposing effective sanctions, the composition and terms of reference of the decision-making organs may receive even more careful scrutiny than in a situation in which their ability to command significant sanctions seems remote. That is, no state is likely to expose itself to the risk of having effective sanctions imposed on it by a body whose membership is inherently hostile to it or that it otherwise mistrusts.

One caution may, however, be appropriate. As the international facilities are likely to be controlled by some intergovernmental organ, the question may arise whether that organ might not itself be an appropriate one to participate in decision-making on sanctions. Such an arrangement would, however, be a mistake, since the interests of the body operating an international facility may, *vis-à-vis* its customers, be different from that of the organ responsible for a nuclear control system. Thus it would seem advisable to keep the latter organ separate from the former. Indeed, as the international facilities should themselves be controlled by some outside authority, the latter might be the same as is charged with responsibility for controlling national activities. On the other hand, it must be clearly provided that sanctions decisions taken by the control authority may not be contravened or even challenged by the organs operating the international facilities that are to constitute the prime devices for the imposition of sanctions.

Imposition of sanctions

As indicated above, although nuclear-related sanctions are the ones that appear to be the most appropriate tools in a nuclear control system, states intent on violating the system are likely to take precautionary measures to shield themselves as far as possible from the impact of such sanctions. In particular, they would attempt to stockpile materials and achieve self-sufficiency in their processing. However, by internationalizing part of the fuel cycle and expecting or especially requiring states to rely on that part of the cycle, any attempts by the latter to achieve nuclear self-suffiency would be negated.

The imposition of sanctions could then be executed simply and neatly by cutting off further supplies from the international facilities to the offending state. Such a cut-off would naturally also include materials owned by the state but previously sent for processing or reprocessing to the facilities in question. The legality of such measures would derive from the consent of the state to its participation in the control system, or from a condition included in the contracts for use of the international facilities, or, in principle, from the Security Council, should it choose to use its power under Articles 39 and 41 of the UN charter to institute a generally binding nuclear control system subject to sanctions.

An appropriate management of the flow of materials to and from the international facilities could enhance the vulnerability of states to the imposition of sanctions, by preventing them from building up stockpiles. Naturally such measures would need to be complemented by the establishment of international storage facilities, as foreseen, for example, in Articles IX.I.1 and XII.A.5 of the IAEA Statute.

Burdensharing

In exercising control in respect of a fully national fuel cycle, a major obstacle in imposing sanctions is the amount and unpredictability of the costs that may result from having to undertake, probably on an *ad hoc* basis, various measures likely to exert sufficient pressure on an offending state. On the other hand, if sanctions can largely be accomplished by merely shutting off the flow of supplies from international facilities, both the amount and the uncertainty in the costs of such sanctions would be moderate.

It should, of course, be noted that the use of international facilities for the purpose of sanctions implementation instead of for the provision of regular flows of materials would not be entirely without costs (e.g., of storing and maintaining of expensive inventories, as well as of losing revenue from the sanctioned state). Although such costs could be borne by the facility, that is, by the states that own or use it, such a distribution is unlikely to be entirely fair or acceptable. Instead in the arrangements that are made for integrating each international facility into the control system, provisions might be included for distributing throughout the entire world community the burden of any sanctions implemented by the facility. The calculation of sanctions costs in a given case is of course a technical accounting matter, but the distribution formula is a political one, in respect of which the competent organs might be guided by the special rules that the IAEA has adopted in relation to the costs of implementing its functions under the NPT (see IAEA General Conference Resolution GC(XX)/RES/341, 1976), or those of the UN in relation to the costs of the Middle-Eastern peace-keeping operations (see, e.g. UN General Assembly Resolution 3101 (XXVIII), 1973).

IV. Conclusions

In deciding whether or not to institute or to participate in efforts to internationalize part of the nuclear fuel cycle, states will of course have to take into account, individually and collectively, a host of complex economic, political and technical factors and to weigh various advantages and disadvantages, benefits and costs, from such an enterprise. In doing so, those states genuinely in favour of effective international controls should consider that an important argument for the establishment of international facilities, is that such facilities will make it far simpler, cheaper and more effective to impose sanctions on recalcitrant participants in the control system—in fact, that internationalization might be considered both a necessary and a sufficient condition for the imposition of sanctions. On the other hand, it is likely that those states whose commitment to non-proliferation is not as firm may be reluctant to join and rely on international facilities through which their own behaviour can more effectively be controlled. Such states might, however, consider that if their own reluctance to yield to effective controls derives from uncertainties about what their neighbours might do, then the dependence of the latter on the same international facilities should do much to enhance the general feeling of security in their region and eventually in the world as a whole.

References

Chayes, A., 1972. An inquiry into the workings of arms control agreements, *Harvard Law Review*, 85:905–69.

Szasz, P., 1979. Sanctions and international nuclear controls, *Connecticut Law Review*, 11(3):545–81.

Szegilongi, E., 1978. Unilateral revisions of international nuclear supply arrangements, *International Lawyer*, 12:857–62.

Paper 21. Internationalizing the fuel cycle: the potential role of international organizations

K.H. LARSON*

Legal Division, International Atomic Energy Agency, Vienna International Center, A-1400 Vienna, Austria

I. Introduction: concerns of nuclear industry

Collapse of the nuclear dialogue

Prior to the initiation of the International Fuel Cycle Evaluation (INFCE), the world nuclear industry was destabilized by commercial and political developments which not only created divisions between supplier and consumer states but also gave rise to differing policies among supplier states. There was increased concern among some states about the risks of nuclear weapon proliferation. In the United States, for example, Congress began a renewed consideration of such risks, culminating in the Nuclear Non-Proliferation Act of 1978, which set forth a series of requirements for the export and use of nuclear material, equipment, facilities and technology of US origin. At the same time there were growing uncertainties in the stability of markets for the supply of nuclear material, services and equipment. Many countries believed that unilateral development of nuclear power policies by supplier states served to erode the existing international régime on the peaceful use of nuclear energy. Some supplier countries believed, however, that unilateral action was necessary to strengthen the existing non-proliferation régime. The nuclear industry was struggling with new governmental policies that were emerging, and the dialogue between countries collapsed in several areas which were thought to be of vital importance for the security of future nuclear power development.

Re-establishing the dialogue

In this atmosphere and in an attempt to create a dialogue which would

*The opinions expressed in this paper are those of the author and do not necessarily reflect the official policy of her professional affiliation.

evaluate the nuclear fuel cycle, the Organizing Conference of INFCE was convened in the autumn of 1977 on a US initiative.[1]

Eight Working Groups were established to examine, in connection with their assigned subjects, the kinds of institutional arrangement that could be part of the future development of the nuclear fuel cycle.

A consensus developed among the participants that although technological developments may help to achieve non-proliferation objectives, the more promising area for further action lies in the development of institutional arrangements.

Perhaps the most important result of INFCE has been the emergence of a political commitment, among a majority of participants, to continue work towards an international consensus to ensure the peaceful uses of nuclear energy. Although much may be accomplished on a bilateral basis at certain stages, there seems also to be agreement that multinational measures may be needed and that consultations must continue in multilateral forums.

Understandably, much of the future study and action with regard to institutional arrangements lies in the political realm and will be the subject for discussion by states. As these countries search for ways to contribute to the post-INFCE period, governments are looking at future institutional arrangements and their relationship to existing international organizations.[2] Because of this, it is necessary to explore the possible nature of that role, particularly with regard to the International Atomic Energy Agency (IAEA). In this paper, 'institutional arrangements' are taken to include the full range of undertakings by either governments or private entities to facilitate the efficient and secure functioning of the nuclear fuel cycle and encompass a broad spectrum of activities, both formal and informal, such as those embodied in commercial contracts, intergovernmental agreements, technical assistance programmes, international studies, safeguards agreements, and multinational and international institutions.[3]

---

[1] The Organizing Conference, attended by 40 countries and 4 international organizations, agreed that the INFCE study was to be technical and analytical and not a negotiation. The Washington Communiqué issued by that Conference established three basic considerations for the study: (*a*) that nuclear energy for peaceful purposes should be made widely available to meet the world's energy needs; (*b*) that effective measures be taken at the national and international levels to minimize the danger of proliferation of nuclear weapons without jeopardizing energy supplies or the development of nuclear energy for peaceful purposes, and (*c*) that special consideration be given to the specific needs of the developing countries.

[2] The term 'international organization' is taken to include regional integration organizations like Euratom, as well as multinational, supra-national or intergovernmental organizations.

[3] Much of the following section is drawn from Section 1 of IAEA (1967).

II. *Development of institutional arrangements*

A historical review

The past 30 years have seen a considerable growth in the number and variety of international institutional arrangements in many fields, often designed to provide a permanent framework for international co-operation, such as that of the specialized agencies of the United Nations. Some of the most successful projects involving international co-operation, however, have been characterized by the pursuit of narrow and clearly defined objectives, involving the participation of only a few countries with common interests and usually in areas where the development of a mature commercial market lay in the distant future. Arrangements of this type in the nuclear area have included Eurochemic [4] and JET.[5] New types of international arrangements, such as Urenco[6] and Eurodif,[7] have recently been formed to meet market demands for enrichment.

New institutional measures to strengthen the nuclear energy régime are likely to emphasize the parallel development of supply assurances and non-proliferation objectives. The development of such measures will probably be achieved on a step-by-step basis and may require a variety of international multilateral and bilateral initiatives. Some issues may lend themselves to solutions in the form of broadly based arrangements for international co-operation. Other issues seem more likely to be solved through more modest means, including informally co-ordinated modifications of government policies or commercial practices. In either case, some problems, such as the assurance of long-term supply, may require a variety of arrangements for their solution.

Future arrangements

Before turning to the potential role of internationalizations, it may be useful to examine the type of institutional arrangement which may be developed in the future.

---

[4] Eurochemic was created under the auspices of the Nuclear Energy Agency of the OECD by 13 states to operate a small reprocessing plant primarily for training and research and development purposes.

[5] Joint European Torus (JET) is a multinational research and development project now being established to investigate fusion energy. See *Official Journal of the European Communities* (1978).

[6] Urenco was created by trilateral agreement among FR Germany, the Netherlands and the UK to co-ordinate previously independent development programmes in centrifuge technology.

[7] Eurodif is a joint stock company incorporated under French law with foreign participation by some states, created to operate a commercial gaseous diffusion plant. For a more detailed description see IAEA (1977). See also OECD (1963) and Urenco (1970).

In recent studies, both within and outside of INFCE, institutional arrangements have generally been examined from the perspective of their potential contribution to non-proliferation and assurances of supply. A lesser degree of consideration has also been given to possible safety and economic benefits and to health and environmental questions.

Many kinds of arrangement have been considered, ranging from informal mechanisms for consultation to binding international agreements. There appears to be increased support for multinational or international arrangements involving both governments and entities of their nuclear industries. These could be applied in the commercial sector, primarily in the form of joint ventures, and through governmental initiatives, such as arrangements for an international plutonium storage scheme and the harmonization of standards and practices throughout all stages of the fuel cycle.

It has been widely agreed that conditions for the establishment of multinational or international institutional arrangements should include membership on a non-discriminatory basis, the application of IAEA safeguards, adequate levels of physical protection for nuclear materials and facilities, means for dispute settlement, and a clear definition of the rights and obligations of the parties.

Harmonization of practices

Growing importance has been attached to international co-operation in the development of recommendations, guidelines and codes of conduct to standardize procedures and harmonize practices in all steps in the fuel cycle.

With regard to assurances of supply, it has been suggested that codes of conduct that regulated relationships with suppliers in regard to the transfer of nuclear materials, facilities and technology and that provided assistance for training and research would be a particularly useful way of developing confidence among consumer countries.

It has also been suggested that national regulations for storage and transport of spent fuel and for waste management be harmonized with a view to the formulation of internationally agreed codes of conduct.

Commercial joint ventures

It has been noted that there may be increased opportunity for, and value in, multinational commercial joint ventures. Studies of the uranium market indicate that investment by consumers in uranium production activities is becoming more common and that many countries are expected to support this trend because it is expected to increase confidence in market stability and lead to reasonable but remunerative prices.

Consumer investment in enrichment facilities through the creation of joint international ventures and the pooling of financial, technical and management resources might contribute to assurances of supplies and markets. In addition, a variety of financial pooling arrangements might make enrichment services more accessible to developing countries.

Fuel cycle management

It has been suggested that important benefits, particularly with regard to non-proliferation, supply assurances and economic considerations, could be obtained from the establishment of facilities under multinational or international auspices. This could involve multilateral participation in a single facility or arrangement or could extend to co-location of facilities in various configurations, including facilities for complete fuel cycle services. Both the reduction of proliferation risks and the increase in supply assurances would depend upon the degree of control exercised by each party to a scheme. Because new arrangements could be time-consuming and difficult to negotiate, national facilities offering services to foreign customers under a common code of conduct might be a useful transitional arrangement if this further step is really needed.

It is also thought that an international scheme for plutonium storage could have enough non-proliferation advantages to justify its establishment. While spent fuel storage presents less proliferation risks than does storage of separated plutonium, international spent fuel management schemes might make more storage space available and thereby provide assistance to certain countries in the cost and management aspects of spent fuel. It might also remove a current political and public relations obstacle to nuclear power.

Multinational and international waste repositories might offer economic, technical and non-proliferation benefits and could be particularly useful in alleviating the concerns of countries with smaller nuclear power programmes. It has been recommended that specific proposals for their establishment should be further elaborated.

Fuel supply arrangements

Working Group 3 of INFCE considered that the availability to some consumer countries of alternative sources of supply or their capacities to stockpile uranium fuel nationally may not be sufficient in the event of major supply interruptions; moreover, countries with smaller nuclear programmes or limited resources find it more difficult and expensive to gain access to commercial fuel-assurance arrangements or to build up adequate stockpiles.

Thus back-up arrangements such as a uranium emergency safety network or an international fuel bank have been considered as alternative short- to medium-term mechanisms. A uranium emergency safety network, as discussed in INFCE, would have both governmental and commercial involvement and would supply fuel (natural and enriched uranium) in emergency situations on agreed terms and conditions to participating states or utilities. Participating states and utilities would contribute to network stockpiles on a national and possibly regional or world-wide basis, and special provision could be made to assist states with new or smaller nuclear power programmes without stockpiles. The fuel bank, using a different administrative structure, would hold assets of both natural and enriched uranium and would provide fuel to participating consumers under agreed terms and conditions in the event of contract default by a supplier which is not the result of a breach of non-proliferation undertakings.

International treaties and instruments

It has been suggested that priority should be assigned to seeking consensus on the conditions for the supply of nuclear materials, equipment, facilities, services and their related technology. Such a consensus could contribute significantly to the assurance of supply as well as to non-proliferation objectives. One product of such a consensus would be the conclusion of an international agreement on the peaceful uses of nuclear energy. Such an agreement would present a visible integrated supply-assurance/non-proliferation régime, but a comprehensive agreement is likely to be a goal which can be reached only after long discussion. It is nevertheless a proposal which is worthwhile and might serve as a goal for the future.

III. The potential role of international organizations

In practice the role of international organizations may prove to be a useful one, but it is essential that both international organizations and their member states analyse the future activities that may realistically be carried out by such organizations.

Some criteria are necessary for this analysis. A useful starting-point might be the following questions:

1. Are the objectives of the proposed arrangement consistent with the objectives of the organization?
2. Is the proposed arrangement of a manageable size?
3. Can it be carried out at an acceptable cost through an acceptable form of funding?
4. Will the organization have a defined set of rules under which to implement the arrangement?

5. Does the proposed arrangement conflict with activities already undertaken by the organization?

Thus the purpose of an arrangement is fundamental in choosing whether to develop that arrangement within an international organization. Certain types of process, such as fuel fabrication, are seen as particularly commercial enterprises, in which an international, intergovernmental organization would not play a major role.

It is widely acknowledged that the development of new institutions or institutional arrangements can be a lengthy and expensive process. In the past, major international institutions generally have taken several years to establish. Even in the case of smaller projects involving international co-operation, implementation of the project may take two or three years following the conclusion of intergovernmental agreements.

In addition, some international projects have not fully met the objectives of the participants. Thus, there is understandable concern that any proposal for new institutions or arrangements be carefully evaluated. Moreover, it has been suggested that new arrangements might be put into effect by modifying existing institutions or by placing new arrangements within existing organizations rather than by setting up new institutional or organizational structures.

However, there is also some hesitation about developing institutional arrangements within existing organizations. It has been argued that they are inefficient and unwieldy and that they are unsuitable because of the nature of their membership and the amount of control over projects exercised by their governing bodies. Another criticism is that commercial confidentiality cannot easily be guaranteed because of their international character. Consequently, it has been suggested that many legal requirements to establish a project are difficult to meet within the framework of existing organizations.

However, one should note that to the extent that these concerns are valid, they will also certainly apply to the creation of new institutions and arrangements. For example, even if the process for the establishment of an IAEA project may be difficult, it is likely that the process would be more difficult if it proceeded on the basis of independent arrangements between the parties, which would in any event require at least a safeguards agreement with the IAEA.

None of the foregoing should be taken to imply that the creation of new, independent institutions and arrangements cannot be justified; rather it is only to suggest that difficulties which exist in present arrangements for international co-operation may be duplicated in new arrangements unless the issues are clearly recognized and analysed from the outset.

Another point merits careful study. One hallmark of the new approach to institutional arrangements is that they are increasingly viewed as interlocking pieces of the same puzzle which can be completed only by designing those pieces over a period of time in relation to the development and design of other pieces. Therefore the process is both evolutionary and integrated. Large schemes are not expected to fall in place immediately, but the isolated development of a single type of arrangement is inhibited. Thus an important question becomes: Under what circumstances can a particular institutional activity be best carried out under the aegis of an existing organization?

A possible role for the IAEA

Certain components of the new nuclear energy régime may be suitable for development within international organizations such as the IAEA. As noted above, supply-assurance issues have been coupled with non-proliferation objectives; this coupling forms the basis for new arrangements that should meet concerns of countries with diverse interests. Involvement of organizations like the IAEA might be particularly useful because of the need to reach consensus among a diverse group of countries.

The IAEA is composed of member states from all the 'nuclear categories' as well as from every geographic region. Its Statute specifically precludes the establishment of programmes which discriminate between member states (IAEA Statute, Article III (C)).[8] In the nuclear area it is the most broadly based existing international forum. In addition, the Statute of the IAEA already provides a basis for the development of a number of activities.

Article III outlines the Agency's primary functions. Under III.A.1, the IAEA is authorized "to encourage and assist research on, and development and practical application of, atomic energy for peaceful uses throughout the world". It may act as an intermediary for the performance of services or for the supply of materials, equipment and facilities by one IAEA member state for another. The Agency may grant certain types of technical assistance by fostering and encouraging the exchange of scientific and technical information and the exchange and training of scientists and experts as provided for in Article III.A.3 and 4.

Under Article III.A.5 the IAEA is authorized to establish and administer safeguards on special fissionable materials and other materials, services, equipment, facilities and information made available by the Agency or under its supervision or control and to apply safeguards at the request of the parties to any bilateral or multilateral arrangement or at the unilateral request of the state.

Under Article III.A.7 the IAEA is authorized to establish or acquire any facilities or equipment useful in carrying out its authorized functions, if the facilities or equipment otherwise available to it are inadequate or

---

[8]This should not be taken to mean that programmes cannot be established which are of interest only to a group of IAEA member states.

available only on terms deemed unsatisfactory by the Agency.[9]

Given this foundation and the Agency's experience, it would seem that the IAEA could be best used in two ways, firstly, to provide a forum for discussion and work by states, which may lead to institutional arrangements that do not necessarily foresee a role for the Agency; and secondly, to assign new functions to the Agency as part of its programme.

A forum for discussion

The first category is the more difficult of the two to describe. One example is the recently concluded negotiation of a Convention on the Physical Protection of Nuclear Materials,[10] which was conducted by member states under the auspices of the Agency. Parties to this Convention, the first international agreement dealing with the physical protection of nuclear material, undertake *inter alia* to apply a certain level of physical protection to nuclear material while it is in international transit. The negotiation of the Convention was preceded by the development, on an international basis, of recommendations and guidelines for the physical protection of nuclear material (IAEA INFCIRC/225/Rev.1 and INFCIRC/254). The IAEA will not, however, play a major role in implementing the Convention.

Similarly, the IAEA might provide a suitable forum for studies and developments in other areas. One possible arrangement which has been discussed recently is the establishment of a forum, following INFCE, to continue the study of important issues. It has been suggested that countries might wish to continue consultations on a multilateral basis. This does not imply the continuation of a major effort like INFCE, but rather a more selective study of certain issues which could be narrowly or broadly defined, depending upon the interests of participating states.

In order to supplement the existing non-proliferation arrangements and increase supply assurances, detailed and internationally agreed measures are needed. These measures must reflect the evolving situation of governmental policies and commercial developments and would accordingly have to be developed in a continuing, step-by-step process. As several groups in INFCE have noted, a major concern is to find ways in which changes in conditions and nuclear policy, particularly in relation to non-proliferation and nuclear supply, can be accommodated on a negotiated

[9] The Statute itself does not establish priorities among these functions. The negotiating history of the Statute indicates, however, that the early expectations of the Agency's founders were that the IAEA would be heavily engaged in activities relating to the storage and supply of nuclear materials, leaving the other functions to play a supporting or secondary role (Szasz, 1970).

[10] The Final Act of the Meeting of Governmental Representatives to Draft a Convention on the Physical Protection of Nuclear Material was signed on 26 October 1979, after more than two years of negotiation. The Convention will be opened for signature at the Headquarters of the United Nations in New York and the Headquarters of the IAEA in Vienna on 3 March 1980.

and agreed basis. A Consultative Committee or Group might conveniently be established under the aegis and using the services of an organization like the IAEA. One of the attractions of such a committee is that its terms of reference would not require it to focus on only one area. Rather it could, if the participants chose, assume a role in co-ordinating the consideration of issues which could then be transferred to negotiating forums or expert groups. This committee could obviously be established outside of the IAEA; however, the Agency might be a more suitable forum, given its representative membership and role in the nuclear industry.

New functions in the IAEA programme

Other types of arrangement may be usefully incorporated into the programme of the Agency. It is recognized that the standardization of practices can contribute to non-proliferation goals while encouraging the reliable functioning of the nuclear industry. International codes of practice have been developed by the IAEA in a number of fields including state systems of accounting, health, safety, transport, regulatory organization, management and disposal of some waste categories, quality assurance and reactor siting, design and operation. It is the Secretariat's experience that developing and achieving widespread adherence to guidelines or a code of practice have proved important in facilitating and strengthening international nuclear co-operation and that such standardization might profitably be extended to a number of other areas, including legal and institutional arrangements.

In addition to current programmes in health, safety and the environment, special attention might be given to the development of ways to improve nuclear supply assurances. As noted above, certain brokerage or supply functions are foreseen by the Agency Statute.[11] Of the two specific proposals discussed by INFCE—the fuel bank and the emergency supply network (see discussion above under *Fuel supply arrangements* in Section II)—it would appear that the IAEA could play a role in the establishment of a fuel bank if states determine that the bank could provide increased fuel assurances, particularly to developing countries or countries with smaller, limited nuclear programmes.

There has been considerable discussion about the role of international organizations in arrangements for fuel cycle management. The IAEA study on regional fuel cycle centres (IAEA, 1977) provided a comprehensive initial evaluation of many aspects of the development of such arrangements. The study has led countries to undertake similar studies specific to their particular situations and needs. As a result there has been an increase in both commercial and governmental interest in the establishment and operation of multinational facilities. Unlike other institutional arrangements,

[11] See Articles IX and XI of the IAEA Statute.

however, organizations like the IAEA are not likely to play a major role in much of this development. For the enrichment and reprocessing stages of the fuel cycle, the emphasis is likely to be on commercial joint ventures established and operated by a small group of countries. The IAEA would, of course, carry out safeguarding activities. In addition, if the consortium of countries were composed primarily of developing countries, then the Agency might also provide technical assistance on some phases of the construction and operation of the facility.

There are, however, certain types of arrangement for fuel cycle management in which organizations like the IAEA might play a role. The IAEA is currently engaged in the study of international plutonium storage and international spent fuel management. The specific basis for the storage of plutonium under the auspices of the Agency is found in Article XII.A.5,[12] while Article III provides the general basis for any scheme in spent fuel management. Although the Agency began the study of these issues in tandem, they have developed into different types of programmes. The Expert Group on International Plutonium Storage is beginning to examine institutional models, while the Expert Group on Spent Fuel is primarily examining ways to improve spent fuel storage techniques economically.

There may also be a role for the IAEA in formulating special arrangements for developing countries. These countries often require special assistance in developing a nuclear power capacity and indeed in the associated infrastructural capabilities. Although bilateral arrangements have been responsive to these needs to a limited extent, it has been suggested that there is a need for the development of international mechanisms to assist in energy planning, financing, manpower development, improvement in industrial infrastructures, and provisions for the transfer of technology, consistent with non-proliferation commitments. It has been suggested that one possible new mechanism that might be established in the setting up of an international technology centre, under the auspices of an existing organization and possibly using existing facilities, to provide on-the-job training and access to expert advice.

IV. Continuing the dialogue

International organizations are themselves the product of international co-

[12] Provisions of this Article form the basis of the IAEA safeguards system today, and Article XII.A.5 contains the as yet unimplemented provision under which the Agency has the right to require the deposit "with the Agency of any excess of any special fissionable materials recovered or produced as a by-product over what is needed...in order to prevent stockpiling of these materials, provided that thereafter at the request of the member or members concerned special fissionable materials so deposited with the Agency shall be returned promptly to the member or members concerned for use".

operation in their establishment and management. Today, they occupy a difficult position. They are often entrusted with responsibilities by their member states; at the same time member states often lack confidence in the ability of the organization to fulfil those responsibilities. This situation can be avoided if states carefully evaluate the capabilities of organizations. There is a role for international organizations in the continuing development of the nuclear energy régime. This role need not be highly political and should not be overestimated. Grandiose plans for international organizations lead to disappointments. Such organizations are not suitable for every institutional arrangement. They can, however, contribute to the continuation of the dialogue between states. It is this dialogue which will lead to consensus and the return of stability to the nuclear régime.

References

IAEA (International Atomic Energy Agency), 1967. Concepts for Institutional Arrangements for the Nuclear Fuel Cycle, Co-chairman/WG.4/67 (A,B), Legal Division of the IAEA, paper submitted to INFCE.

IAEA (International Atomic Energy Agency), 1977. *Regional Nuclear Fuel Cycle Centers*, vol.2 (IAEA, Vienna), pp.66–69.

OECD (Organisation for Economic Co-operation and Development), 1963. Statute of the European Company for the Chemical Reprocessing of Irradiated Fuels (Eurochemic), OECD Publication No. 15979.

Official Journal of the European Communities, 1978. Council decisions of 30 May 1978 on the establishment of the "Joint European Torus (JET) joint undertakings" in issue of 7 June 1978, *L151/8*.

Szasz, P., 1970. The law and practices of the International Atomic Energy Agency, *Legal Series No.7* (IAEA 351, 352).

Urenco, 1970. Agreement on Collaboration in the Development and Exploitation of the Gas Centrifuge Process for Producing Enriched Uranium, Urenco Paper 4315 (Her Majesty's Stationery Office, London).

Index